Theoretical Alchemy

Modeling Matter

T0275915

Theoretical Alchemy
Modeling Matter

Walter Harrison
Stanford University, USA

 World Scientific

NEW JERSEY · LONDON · SINGAPORE · BEIJING · SHANGHAI · HONG KONG · TAIPEI · CHENNAI

Published by

World Scientific Publishing Co. Pte. Ltd.

5 Toh Tuck Link, Singapore 596224

USA office: 27 Warren Street, Suite 401-402, Hackensack, NJ 07601

UK office: 57 Shelton Street, Covent Garden, London WC2H 9HE

British Library Cataloguing-in-Publication Data
A catalogue record for this book is available from the British Library.

THEORETICAL ALCHEMY
Modeling Matter

ISBN-13 978-981-4322-13-3
ISBN-13 978-981-4322-14-0 (pbk)

Printed in Singapore by Mainland Press Pte Ltd.

To Lucky, and to our delightful and ever-growing
extended family

spirit

Author's Note

If this book has a thesis, it is that the best way to understand chemical bonding may be to take the appropriate view for each molecule or crystal, a view which may be quite different for different systems. Sometimes two very different views are appropriate for the same system, and then the combination may even give the parameters needed to estimate properties such as cohesion. One then proceeds by hand to make the estimate. This way of proceeding contrasts strongly with Density Functional Theory, certainly the method of choice for most such studies involving solids. In that approach one seeks a single view, which may be applied to *all* systems, and then approaches it computationally. Perhaps that is another example of a different view also being appropriate.

The book might also be described as "everything you need to know in order to understand oxide fuel cells", since that is the center of my current research. However, that understanding requires an extraordinarily large background. The fuel cells show up particularly in the last section of the last chapter where oxygen adsorption on manganites is discussed, and comparison is made with current Density-Functional treatments. In this regard, the book is an argument for a different way of proceeding with such problems from the mainstream approach.

This book also contrasts with two previous books. The first, *Electronic Structure and the Properties of Solids*, appeared in 1989. In the midst of writing that text a set of empirical rules for obtaining coupling between neighboring atomic states, such as the coupling between neighboring atomic s states as $-1.42 \, \hbar^2/md^2$, became a theoretical finding, $-9\pi^2/64 \, \hbar^2/md^2$. There was time to reorganize for that and I was lucky enough to be able to redo the entire program ten years later in *Elementary Electronic Structure*. Both texts were somewhat massive, and I am delighted to have the chance ten years later to try to distil the essence of the representation of electronic structure in a much briefer description.

The two earlier books were written as if the reader could follow each point; the present one more nearly envisages skimming, with the hope that something interesting or useful will be found. To be sure, the new one centers on the insights described in the earlier books, and a series of my subsequent papers. It does not give due credit to the essential contributions of the many other workers who made these insights possible, and who are referenced in the two earlier books. This book might otherwise have grown to a similar size. What references there are, were mostly included to indicate where more complete descriptions exist than I could include here.

The book is also made briefer by focusing primarily on the bonding energies, the energy gained in assembling a molecule or a solid, or a solid with a surface. A central point is that the same description of the electronic structure which gives the cohesion can also be used to understand all of the other properties, but we largely leave that to the other books. We also leave any extended derivations to the earlier texts except where they are new, particularly in the extension to molecules, and then they appear here in the Appendixes. However, the subject is quantitative and we use equations as an integral part of the discussion.

A note also on the typography, which may leave something to be desired. I have prepared camera-ready copy using Microsoft Word, in the interests of economy. All shortcomings can be attributed to me, and not to World Scientific Publishing Company. They have my gratitude for providing books at reasonable prices, hardly the current fashion. Perhaps the reader can accept the result as something between a personal letter and a traditional book.

One particular view which has proven helpful is "theoretical alchemy" in which we understand one simple element, like the inert gas argon, and then conceptually remove a proton from the nucleus of one atom, making it a chlorine nucleus, and place it in the nucleus of another argon atom, making it a potassium. The electron cloud will shrink a little around the potassium and expand around the chlorine, but we can understand the behavior of the potassium chloride compound in terms of the simpler inert gas. It is not the central idea of this book, but it comes up enough to seem an appropriate title, rather than something like "Reinventing Chemistry". We also like the combination of the modern "theory" with the ancient effort of "alchemy" to make sense of the material world. A goal of the original alchemy, transmutation of lead to gold, is theoretically quite easy; one simply removes three protons from the nucleus of lead and it is done. We leave it to our experimental colleagues to work out the details in the laboratory.

Walter A. Harrison
April, 2010
Stanford, California

Note about the cover: The painting, by the author, intends to represent this book. The alchemical symbols below are identified at the beginning of each chapter. The red symbol inside the flask is for oil of vitriol, the blue symbol is ammonia; think of litmus paper. The corresponding reaction is given in the lower part of the paper. There is also a faint symbol in the flask for a hydrogen bond in water, from Chapter 2, and Appendix 2C on the nature of the hydrogen bond. At the top of the paper is an equation from the electronic structure of semiconductors in Chapter 5. We hope the book, like the candle, can shed some light on a modern way of understanding matter.

Contents

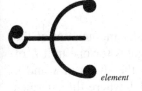

element

CHAPTER 1

Atomic States

An understanding of electronic structure would seem at the outset to be an impossibly difficult task. Even in a small atom, the several electrons strongly interact with each other as well as with the nucleus, presenting an intractable dynamical problem both classically and in quantum mechanics. The only conceivable hope would be that it was possible to consider each electron by itself, in the presence of some average effect from the others. If that were somehow possible, then in quantum mechanics the wavefunction representing the many electrons, $\Psi(\mathbf{r}_1,\mathbf{r}_2,\mathbf{r}_3,...)$ could be written as a product of one-electron functions $\psi(\mathbf{r}_1)\psi(\mathbf{r}_2)\psi(\mathbf{r}_3)....$ D. R. Hartree (1928) first gave that a try. Inserting this speculated form in the Schroedinger Equation he used a variational calculation to derive the self-consistent one-electron equations which have provided the basis for virtually all subsequent studies of atoms, molecules, and solids.

The energy levels of individual atoms for each of the elements are absolutely central to our effort to understate the bonding energy for all types of systems. We assume a basic understanding of quantum mechanics, though only a small part of it will be necessary to our discussions. A brief introduction to the needed parts was given in *Elementary Electronic Structure*, Harrison (1999), and a more complete presentation in *Applied Quantum Mechanics*, Harrison (2000), and of course many other places.

1.1 Atomic Energy Levels

The atomic states are of course classified by the angular momentum of the electron, s states for zero angular momentum, and p states for one unit $\hbar$ of angular momentum. The row number n distinguishes the shells of s and p states of increasing energy, as shown here in Table 1.1 where the energies are given for the states in the shell being filled. It is a very familiar fact that after filling the $1s$ state for hydrogen with one electron and helium with one electron of each spin, the successive filling of the ns state and the three np states, with electron of both spins, provides the nth row of the periodic table, with the electrons in the nth shell, being filled, called valence electrons. Here the elements to the right of the inert gases, Columns IA and IIA, would be shifted down one row and to the left to fit the numbering scheme of the rows, but we shall seldom use the n and we shall see that this arrangement is convenient. There are also series of transition metals with partly filled d shells, each with two units of angular momentum, to the left of each element in Column IB, and partly filled f shells, with three units of angular momentum, to the right of barium and radium in Column IIA; we return to those in Chapter 7.

For each element we have given in Table 1.1 the energies of the highest-energy occupied s and p states, obtained from Hartree-Fock calculations by Mann (1967). [In the column IIB the p state is not yet occupied, but the energy was estimated from surrounding values. In columns IA and IIA the p state energies are from the previous row, therefore core-state energies.] We may think of these energies as the energy required to remove an electron from the corresponding state in the neutral atom.

The systematics of these values is interesting. The energies of each category become increasingly negative, approximately in proportion to row number, across the table, with the s state approximately twice as deep as the p state. [Roughly, ε_s is Column number times the energy of the hydrogen $2s$ state, -3.4 eV.] This also includes the IA and IIA columns which appear to the left in the more usual arrangements of the periodic table. The energies, by column, are also approximately the same in each row, the basis of the periodic system, though they are slightly deeper in the top row, beginning with beryllium.

We shall often need to designate these states in diagrams, and we do that as indicated in Fig. 1.1. The states are denoted by traditional *kets*, |s> for the s state, etc. The |s> is spherically symmetric, so drawn as a circle. The

Table 1.1. Hartree-Fock term values for valence levels in eV, from Mann (1967). The first entry is ε_s; the second is ε_p (values in parentheses are highest core level). The third entry is the empty-core radius, r_c in Å. The fourth is U in eV, the intra-atomic Coulomb repulsion between electrons, from Harrison (1999).

IB	IIB	III	IV	V	VI	VII	VIII	IA	IIA
						H	He	Li	
						−13.61	−24.98	−5.34	
								-	
								0.92	
								8.17	
	Be	B	C	N	O	F	Ne	Na	
	−8.42	−13.46	−19.38	−26.22	−34.02	−42.79	−52.53	−4.96	
	−5.81*	−8.43	−11.07	−13.84	−16.77	−19.87	−23.14	(−41.31)	
	0.58	-	-	-	-	-	-	0.96	
	10.25	10.26	11.76	13.15	14.47	15.75	15.00	6.17	
	Mg	Al	Si	P	S	Cl	Ar	K	Ca
	−6.89	−10.71	−14.79	−19.22	−24.02	−29.20	−34.76	−4.01	−5.32
	−3.79	−5.71	−7.59	−9.54	−11.60	−13.78	−16.08	(−25.97)	(−36.5)
	0.74	-	-	-	-	-	-	1.20	0.90
	7.28	6.63	7.64	8.57	9.45	10.30	11.12	5.56	6.40
Cu	Zn	Ga	Ge	As	Se	Br	Kr	Rb	Sr
−6.49	−7.96	−11.55	−15.16	−18.92	−22.86	−27.01	−31.37	−3.75	−4.86
−3.31	−3.98	−5.67	−7.33	−8.98	−10.68	−12.44	−14.26	(−22.04)	(−29.88)
-	0.59	0.59	-	-	-	-	-	1.38	1.34
7.07	7.83	6.61	7.51	8.31	9.07	9.78	10.48	5.02	5.71
Ag	Cd	In	Sn	Sb	Te	I	Xe	Cs	Ba
−5.99	−7.21	−10.14	−13.04	−16.03	−19.12	−22.34	−25.70	−3.37	−4.29
−3.29	−3.89	−5.37	−6.76	−8.14	−9.54	−10.97	−12.44	(−18.60)	(24.60)
-	0.65	0.63	0.59	-	-	-	-	1.55	1.60
6.34	6.95	6.00	6.73	7.39	8.00	8.58	9.13	5.05	5.70
Au	Hg	Tl	Pb	Bi	Po	At	Rn	Fr	Ra
−6.01	−7.10	−9.83	−12.49	−15.19	−17.97	−20.83	−23.78	−3.21	−4.05
−3.31	−3.83	−5.24	−6.53	−7.79	−9.05	−10.34	−11.65	(−17.10)	(−22.31)
-	0.66	0.60	0.57	-	-	-	-	-	-
6.75	7.33	6.30	7.03	7.68	8.28	8.85	9.39	4.93	5.54

wavefunction for the p state $|px>$ is a spherically symmetric function times a factor x/r, positive for $x > 0$, negative for $x < 0$ and equal to zero in the yz plane. $|py>$ and $|pz>$ are similarly defined. We may think of s and p as standing for "spherical" and "polar", though they really relate to the spectra, "sharp", and "principal". One d state is also shown.

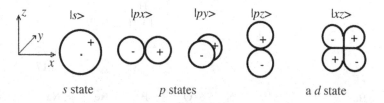

Fig. 1.1. The symbols which are used to represent the s, p, and d states of the atom. "$|s>$" is the Dirac notation for the s state.

1.2 Atomic Levels as a Basis for Other Systems

Just as Hartree tried the product form for the many-electron wavefunctions to obtain one-electron states for individual atoms, we may imagine trying one-electron states for molecules or solids which consist of sums of the atomic states for the atoms which make up the molecule or solid. This is called the *linear-combination-of-atomic-orbitals* (LCAO) method, or the *tight-binding* method when applied as we shall do here, obtaining the needed parameters from other considerations. We shall see how this is carried out explicitly for diatomic molecules in Chapter 3, and how the coupling between neighboring orbitals is obtained when we treat simple metals in Chapter 4.

We may comment on the arrangement of elements in Table 1.1, first suggested by Pantelides and Harrison (1975). As in other arrangements they read like a book in order of atomic number, but the break is at a different point. Column IV is special as containing covalent solids, the elemental semiconductors. Column VIII is also special, the inert gases with full shells of electrons. All of the nonmetals lie between these two columns, at the center of this table. The simple metals lie outside these columns, called type A to the right and type B to the left. Beyond these, and not shown in the table, are transition metals and f-shell metals which we return to in Chapter 7.

The top row with C, N, O, F, and Ne is particularly interesting, and O, at the center, is the star of the show. If we combine a nonmetal with a metal to the left, we obtain a covalent compound, ordinarily in the same diamond structure (which we shall describe in Chapter 5) as are most of the Column IV elements, and semiconducting. The compounds Ge, GaAs, ZnSe, and CuBr are a typical example. If we combine a nonmetal with a metal on the right, we obtain an ionic insulator, ordinarily with the structure of rock salt, and electrically insulating (which we shall describe in Chapter 6). It is interesting that the distinction between these two very different types of compounds comes only in the solid; there is no such fundamental difference between the two diatomic molecules GaAs and RbBr.

A useful way of thinking about these two types of series can be called "theoretical alchemy", in which we switch from element to element by removing protons from a nucleus or adding them. Thus we shall discuss the formation of bonds in germanium in the diamond structure in Chapter 5. With four electrons, it chooses this structure with four neighbors in a tetrahedral arrangement, placing one electron in each of the four two-electron bonds with its neighbors. We can then imagine freezing the electron states and shifting a proton from one Ge nucleus, making it a Ga nucleus, and putting the proton in a neighboring Ge nucleus, making it As. If we unfreeze the electrons, the bonds will shift a little towards the As and we have GaAs. We may repeat the process to obtain successively ZnSe and Cu Br, and the properties change smoothly in the process.

Similarly we may start with a solid made of the inert-gas argon, arranged in a simple-cubic crystal lattice. We may again freeze the electron states and transfer a proton, making Cl on one site and a potassium (K) on the other. In this case it was the element receiving the proton which became a metal, while for the covalent solids it was the element giving up a proton. If we unfreeze the electron states the complete shell of electrons in the chlorine will expand to become a chlorine *ion*, negatively charged, and the shell on the potassium will shrink to become a potassium ion, positively charged. We have constructed the ionic insulator, potassium chloride. We may similarly construct calcium sulphide, or RbBr and SrSe if we start with krypton, but we are not guaranteed that all are stable in nature. In Chapter 6 we see how to estimate cohesion and other properties of these ionic compounds using the energies from Table 1.1.

For most of these solids we will use directly the energies of the atomic states of the constituent atoms listed in Table 1.1. We will also need the coupling between the atomic states on neighboring atoms, which we introduce in Chapter 3 when we discuss diatomic molecules. These energies and these couplings are all we will need to estimate a wide range of properties for both these classes of systems. What approximations are appropriate for each solid depends upon the arrangement of the atoms in the solid, and only secondarily on which elements are involved since that determines the structure. Magnesium could have been placed to the right of sodium, rather than to the left of aluminum. Actually MgS forms in a tetrahedral, diamond-like, structure as well as in a rock-salt structure and is treated quite differently in the two different structures. The hydrides, the compounds of hydrogen discussed in Chapter 2, can be thought of quite differently than one thinks of these covalent and ionic systems.

1.3 Electron Density Distributions

When we treat individual systems we shall sometimes need to know something of the atomic states themselves, as well as the energies, and we should discuss them here. Starting at the top, the wavefunction, or orbital, for the hydrogen $1s$ state of Table 1.1 is given by

$$\psi(r) = \sqrt{\frac{\mu^3}{\pi}} e^{-\mu r} \tag{1.1}$$

with μ related to the energy by

$$\varepsilon_s = -\frac{\hbar^2 \mu^2}{2m}. \tag{1.2}$$

It will be convenient to know that $\hbar^2/m = 7.62$ eV-Å^2, as well as $e^2 = 14.4$ eV-Å. For hydrogen then $\mu = 1.89$ Å^{-1}, the reciprocal of the Bohr radius. For helium the $1s$ state is approximately given by Eq. (1.1), but with $\mu = 2.56$ Å^{-1}, obtained from Eq. (1.2). It is only approximate since with a second electron present the potential energy of an electron is no longer exactly of the form $-e^2/r$ which leads to Eq. (1.1).

Eqs. (1.1) and (1.2) are in fact not bad representations of the wavefunctions for the s states of the heavier elements. They differ strongly only at small distance. It is useful to discuss this in terms of the electron density

$$\rho_0(r) = \psi^*(r)\psi(r) = \psi(r)^2 \approx \mu^3 e^{-2\mu r}/\pi \tag{1.3}$$

associated with such a wavefunction, or the probability $P(r)dr$ that the electron lies in the spherical shell of radius r and thickness dr, given then by

$$P(r) = 4\pi r^2 \rho_0(r) \approx 4\mu^3 r^2 e^{-2\mu r}. \qquad (1.4)$$

The three p states also have angular factors which traditionally are written as the spherical harmonics $Y_\ell^m(\theta,\phi)$, with the radial function written $R(r)$ so that the wavefunction is $\psi(\mathbf{r}) = R(r)Y_\ell^m(\theta,\phi)$. We shall write the angular part in an equivalent form, $x\sqrt{3}/(r\sqrt{4\pi})$, $y\sqrt{3}/(r\sqrt{4\pi})$, and $z\sqrt{3}/(r\sqrt{4\pi})$ for the three p states, which we refer to as the x, y, and z oriented p states. Note the important feature that if we square the three and add them, the resulting charge density is spherically symmetric, and the integral over volume gives three electrons. It turns out that Eqs. (1.1) and (1.2) are also not bad approximations for the $R(r)$ for p states, as well as for s states (for which $R(r) = \psi(r)$), again differing primarily at small r, and the same $P(r)$ of Eq. (1.4) is a reasonable approximation. We may see how well this does by plotting the $P(r)$ obtained from Mann's (1967) wavefunctions for neon, the first inert gas for which both s and p states arise. It is compared with Eq. (1.4), now based on Eq. (1.1) with $\mu = 2.46$ Å^{-1} for the p states and 3.71 Å^{-1} for the s states, in Fig. 1.2.

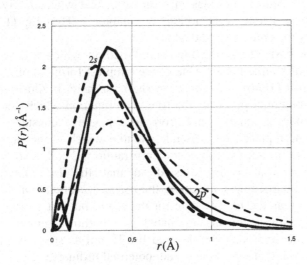

Fig. 1.2. The radial distribution of charge for electrons in neon, heavy lines for valence $2s$ states, and light lines for $2p$ states. The solid lines are from Mann's (1967) Hartree-Fock calculations, the dashed lines from Eq. (1.4).

There are certainly significant differences in the approximate forms, but they are qualitatively the same at large r where they are used in Chapter 2. We shall repeat these curves there when we use them. We could of course adjust μ to obtain a better fit, but we choose not to do that here.

Fig. 1.2 also gives a clue to the origin of the periodicity of the properties of the elements reflected in Table 1.1. The wavefunction for the s state in neon is very similar to that for the s state in helium, just above it in the Table. The μ differs by a factor of 0.7 and it has an extra bump at small r. For argon, below neon, the μ differs by another factor nearer one and there is a second bump at small r, and this continues for the subsequent elements below argon. The properties of the atom are dominated by the outer part of the wavefunction and therefore are very similar for all atoms in the same column of the Periodic Table.

1.4 The Basis of Pseudopotentials

The extra bump r in Fig. 1.2 comes from the extra structure in the $2s$ state, which is necessary to keep it orthogonal to the $1s$ state, and the same is true for a second bump for a $3s$ state. The important fact for us is that this has only a small impact on the normalized wavefunction at large r where it will be used for the central hydrides in Chapter 2, and where it gives rise to coupling between neighboring orbitals in Chapter 3. Thus Eq. (1.3) can be meaningful for the entire periodic table.

This is the basis of the pseudopotential method, which will be central to our treatment of simple metals. One of the simplest formulations, suggested first by Ashcroft (1966) and which we shall use here, is illustrated in Fig. 1.3. The true potential produces the true wavefunction which goes through zero, has a node, at small r and drops to a negative constant. That true potential can be replaced by a pseudopotential which is the same at large distances but set to zero within some core radius r_c, called an *empty-core pseudopotential*. That r_c is adjusted so that the $1s$ solution of the corresponding Schroedinger Equation, the *pseudowavefunction*, would have the same energy as the real $2s$ state for the atom, but not go through zero. Values for the simple metals are listed as the third entry in Table 1.1 Exactly the same approach can be used for $3s$ and $4s$ states in the heavier atoms. It turns out that this same pseudopotential ordinarily gives p states of the right energy for that row, also without the extra structure near the nucleus. Thus, an r_c can be obtained for each element which gives the s-state

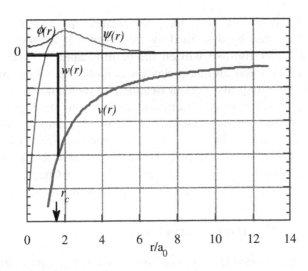

Fig. 1.3. The heavy line $v(r)$ shows the potential for a lithium atom and the thin line $\psi(r)$ shows the corresponding valence 2s-state with energy ε_s. If the potential is replaced by the empty-core pseudopotential $w(r)$, the lowest-energy s-state becomes $\phi(r)$ again with energy ε_s. This $\phi(r)$ is the pseudowavefunction representing the valence state. Radii are shown in units of the Bohr radius a_0. For lithium the core radius is $1.73\ a_0 = 0.92$ Å from Table 1.1.

energy in Table 1.1, and that same pseudopotential is added for all the atoms in a metal crystal. The pseudowavefunctions for the metal then have none of the intricate structure near the nuclei and are much simpler than the true wavefunctions. Again, the only parameters which are needed to understand the electronic structure of the simple metals are the s-state energies given in Table 1.1, which determine the empty-core radius.

The metallic elements form in densely packed structures which we shall describe. Then the superimposed atomic pseudopotentials turn out to be sufficiently weak that the electrons to a good approximation are in free-electron states, only weakly perturbed by the pseudopotential. The true potential is so strong that it makes bound states, the core states, at each atom, but these are eliminated by the pseudopotential, and we can estimate what effects the remaining pseudopotential has by perturbation theory.

The fact that the pseudopotential is weak does not mean that the states in the metal cannot also be suitably represented by linear combinations of the atomic states of Table 1.1 as we suggested at the beginning of Section 1.2.

Quite remarkably, the fact that the states in metals, as well as semiconductors, can be described by these sums of atomic states and are also quite free-electron-like will enable us in Chapter 4 to determine what the couplings between atomic states on neighboring atoms must be, and we use those couplings also for our treatment of molecules.

This completes a brief description of how we may conceive of the electronic structure of systems constructed from the elements listed in Table 1.1. It also indicates how we can make approximations to estimate the properties of all of those systems. The presence of d and f states in other elements creates complications, but we shall see in Chapters 7 and 8 that they can also be dealt with in rather direct and simple ways.

1.5 Electron-Electron Interactions

In spite of the remarkable success of this one-electron approximation, there will be a number of occasions when we must look beyond it. This turns out to be surprisingly simple to do. First, for an isolated atom, the energy levels we have given represent the negative of the energy required to remove an electron from the neutral atom, the ionization energy, $-\varepsilon_s = 5.34$ eV for the electron in lithium (experimentally 5.39 eV for lithium). If we remove an electron and then return it we gain back this same amount of energy. However, if we add an electron to a *neutral* atom — even if it is to the same state (in lithium the s state with opposite spin) — we will gain less energy, reduced by the Coulomb interaction U with the first electron, which is still on the atom. Estimates of that U for each element are the fourth item in Table 1.1. Adding it to the energy level listed in Table 1.1 gives the *electron affinity* of the atom for that state. In lithium it comes out positive, indicating that the lithium ion will not hold a second s electron, but many atoms can. This U will be a useful number in our studies. It will also become important in Section 3.5 to note that when the electron which we add to a neutral atom came from another neutral atom a distance r away, this U is reduced by $-e^2/r$, the Coulomb potential from the charged atom left behind. It is the very near cancellation of these two terms when we form molecules, which often, but not always, makes it unnecessary to worry about either.

A second, more subtle, effect of electron-electron interactions arises if we consider two atoms, not overlapping each other but in the same vicinity, called the *van-der-Waals interaction*, for which we give a derivation in Appendix 1A. This is most usually thought of as arising from zero-point fluctuations in one atom, producing a dipole and therefore an electric field dropping off as $1/r^3$. That field in turn causes a dipole on the second atom, equal to the field times the polarizability α. That induced dipole then gives a field, again proportional to $1/r^3$ interacting with the first zero-point dipole, lowering the energy in proportion to $1/r^6$. Our derivation in Appendix 1A is more direct, but is equivalent to this, leading to a van-der-Waals interaction energy given by

$$E_{vdW} = -\frac{11\alpha^2(\varepsilon_p - \varepsilon_s)}{16r^6},$$ (1.5)

for two atoms, each with one electron in a ground state of energy ε_s but also having empty excited states of the atom of energy ε_p.

In Appendix 1A we also obtain a formula for the polarizability of such an atom, noting that an electric field in a z direction, giving a potential energy for an electron of eEz, will give a coupling between these two electron states of eEP_{sp}, with

$$P_{sp} \equiv <p_z|z|s> \equiv \int \psi_{pz}(\mathbf{r})z\psi_s(\mathbf{r})d^3r.$$ (1.6)

[This definition of a "coupling" between states $|1>$ and $|2>$ as $<1|H|2>$, with H the Hamiltonian or energy operator, will be clarified at the beginning of Chapter 3.] This is evaluated for the model wavefunctions of Eq. (1.2) and applied to the inert-gas atoms of Column VIII in Table 1.1. [In that case the energy of an excited s state, not given in Table 1.1, is needed and taken to be $\frac{1}{2}\,\varepsilon_p$.] It leads to a polarizability of

$$\alpha = 0.94\frac{\hbar^2e^2}{m\varepsilon_p^2}$$ (1.7)

for inert-gas atoms. These are compared with experiment in Table 1.2. The estimates are meaningful, but clearly we have missed the important contributions for the heavier elements, the excited p and d states giving a much larger contribution than our assumed s state (Harrison, 2006b). Similar discrepancies will not be uncommon in our simplified analyses.

We may also use Eq. (1.5) to estimate the cohesion of inert-gas solids. Inert gases condense at low temperatures into close-packed structures with twelve nearest neighbors. The energy gain is six times Eq. (1.5) per atom.

If we assume that the repulsion holding the atoms apart is proportional to the square of the attraction, E_{vdW}, (as we shall use for other systems, and which here becomes the familiar Lennard-Jones interaction), it will cancel half of the gain so that the cohesive energy (taken as negative when binding) is $E_{coh} = 3E_{vdW}$. These are evaluated in Table 1.2, using the experimental polarizability, giving again only qualitative agreement with experiment.

Table 1.2. Polarizabilities and Cohesion for Inert Gas Atoms

Element	$\alpha(\mathring{A}^3)$Eq. (1.7)	$\alpha(\mathring{A}^3)$ Exper.[a]	$E_{coh} = 3E_{vdW}$ (eV)	E_{coh} (eV) Exper.[b]
He	0.17	0.21		
Ne	0.19	0.40	−0.004	−0.020
Ar	0.40	1.66	−0.016	−0.080
Kr	0.51	2.53	−0.023	−0.116
Xe	0.67	4.12	−0.033	−0.16

[a]Holm and Kerl (1990). [b]Kittel (1976).

1.6 Nuclear Structure

For most purposes here, the only properties of the nucleus which will be needed are its mass and its charge. However, the nucleus does have structure on its own and perhaps no book with "alchemy" in its title should go without discussing it. Here we give only the most basic aspects of nuclear structure. More details are available from many sources, such as *Nuclear Structure Theory* by Irvine (1972).

The nucleus is composed of *nucleons*, the positively charged *protons* and the *neutrons*, without charge. Both have a spin angular momentum of $\frac{1}{2}\hbar$, as do electrons, and very nearly the same mass as each other, about 2000 times that of the electron. The small difference is important because the neutron, being slightly heavier and having therefore greater rest energy, can beta-decay (with a half-life of several minutes) into a proton by emitting an electron (beta-ray, or β-ray) and a neutrino. This need not occur in the nucleus where the Coulomb energy from the other protons raises the energy of the proton. They both have magnetic moments but, not surprisingly, they are different. They are held together in the nucleus by strong, short-range forces, which are very much the same between any pair, protons with protons, protons with neutrons, and neutrons with neutrons. Those forces are provided by π-mesons, in the sense that the Coulomb interaction between charges is provided by photons.

The experimental properties of the ground state of the nucleus are very reminiscent of that of a drop of ordinary liquid, on a very much smaller scale;

the diameter of the nucleus as determined by scattering experiments is several times 10^{-13} cm. The corresponding view is called the *liquid-drop model*, discussed in the earliest days by Bohr. These drop-like properties are that the nuclei are approximately spherical and the volume, as measured by scattering experiments, is approximately proportional to the number of nucleons making up the nucleus. Further, the binding energy of the nucleus — the energy required to separate it into individual nucleons — is also approximately proportional to the number of nucleons. Both are properties shared by molecules such as water making up a drop of liquid.

Much more detailed properties of the nucleus can be obtained by taking the same one-particle view which we took above for electrons, called the *shell model* for the nucleus. This approximation can be justified by the same variational calculation which we shall use for electrons in Chapter 3. Then we say that each nucleon moves in the average potential arising from the interaction with all of the other nucleons. This potential should be spherically symmetric for the spherical nucleus and the liquid-drop model would suggest a square-well potential as illustrated in Fig. 1.4. Then the one-particle states for a proton can again be classified by their angular momentum. Further, the protons have half-integral spin and will obey the Pauli Principle, filling the lowest-energy states just as they were filled with electrons in atoms. Since the neutrons have almost the same strong internuclear interaction, the neutron states will have very similar energies, with the shifts due to lack of charge being quite small. The neutrons also obey the Pauli principle, and fill the neutron states independently of the

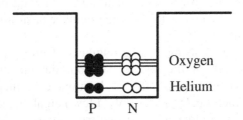

Fig. 1.4. The one-nucleon potential in the shell model is a spherical well of approximately constant depth. The 1s-state is lowest, and can be occupied by two protons and two neutrons, giving the helium nucleus. The next level is a p-state, which can accommodate up to six additional protons and six additional neutrons. Successively adding nucleons to He gives Li, Be, B, C, N, and O. The full shell is O^{16}, with 16 nucleons.

filling of the states by protons since the neutrons and protons are not identical to each other and can both occupy the same states. This is just as electrons — or protons — of different spin can occupy the same orbital.

This shell model immediately suggests some of the most important properties of nuclei. First, as in atoms, there will be shells of levels of increasing energy. If there are two protons in the nucleus, helium, they will occupy the lowest s-state, as in Fig. 1.4. The next highest state in the spherical cavity is a p-state, so in lithium, with three protons, the third will go into that p-state at considerably higher energy. This gives a correction to the liquid-drop model, indicating that the total binding of the nucleons is not *exactly* proportional to the number of nucleons, but a nucleon added to a newly filled shell will be more weakly bound. This happens again with a total of eight protons (oxygen) where the p-shell is filled (Fig. 1.4) and in the fluorine nucleus the additional proton must go into the 2s-state.

At the same time that we are filling the proton states, it will be favorable to fill the neutron states, of very nearly the same energy. If in lithium (three protons) we had not put two neutrons into the neutron 1s-state, this third proton would decay by emitting a positron (the antiparticle version of beta-decay) to transmute the nucleus to helium with an additional neutron. There must always be approximately the same number of protons and neutrons for these light nuclei. In particular, the helium nucleus with two protons and two neutrons filling the 1s-nuclear states is especially stable, as is the oxygen nucleus with eight protons and eight neutrons, both shown in Fig. 1.4. When a nuclear shell of protons is partly filled, the same shell of neutrons can be partly filled with a different number of neutrons without producing the instability mentioned for lithium, allowing different *isotopes* of the same element, nuclei with the same number of protons, but different numbers of neutrons.

As we move to increasingly large numbers of protons and neutrons in the nucleus, the depth of the square well binding the nucleons remains approximately constant because the nucleon-nucleon interaction is of so short a range that each nucleon sees only a few neighbors at one time. Thus the well expands in volume and the one-particle states become closer in energy. As in adding atoms to a metal the Fermi energy remains about the same, as does the cohesive energy per nucleon, and the volume increases, all as suggested by the liquid-drop model. However, as there are more and more protons, the Coulomb interaction between them raises the proton energy more and more above the neutron energy and it becomes favorable to have *more* neutrons than protons, up to 50% more for the heavier elements. Otherwise the protons would emit positrons to produce more neutrons.

This simplest shell model also describes excited states of nuclei, analogous to excited electronic states of atoms. Also as in atoms a nucleus in an excited state can emit a photon and drop to the ground state. These processes can be calculated just as we calculate them for electrons in atoms, molecules, and solids. In the case of nuclei, which are so strongly bound, the energy differences are huge and the photons have energies of the order of millions of electron volts, *gamma rays*, rather than a few electron volts for electronic transitions in atoms.

The shell model provides an understanding of the magnetic moments of nuclei. In particular, in the helium nucleus with the proton 1s-state filled with both spins, and with the neutron 1s-state filled, the nucleus has no net spin and no magnetic moment. The same is true of the nucleus of oxygen with its 2p-shells completely filled. The magnetic moments of *other* nuclei allow nuclear magnetic resonance (NMR) when magnetic fields are applied and microwave radiation is used to cause transitions between different orientations of the nuclear magnetic moment.

The shell model also provides the basis for the theory of fission and fusion of nuclei. It is of course an approximate theory as is our theory of electronic states in atoms, but again a very successful one. For the case of fission of a heavy nucleus, such as uranium with a ratio of neutrons to protons of 1.6, into two lighter nuclei, with smaller ratios for the stable isotopes, it is not surprising that extra neutrons are emitted. These neutrons causing fission of other uranium nuclei is of course the origin of the chain reactions in nuclear reactors and bombs. Much more detailed theory is necessary to describe such processes well. One of the most important refinements of the theory is the addition of spin-orbit coupling, which is present also in atoms, though not discussed here.

There is also structure to the nucleons, each being constructed of three *quarks*, and held together by *gluons*. Indeed the quarks may be without mass, so the nucleon mass arises from the binding of the quarks together. The corresponding Standard Model of fundamental particles is beyond the scope of this book, and of this author. Isolated quarks have not been observed, and indeed they may be unobservable in principle. Just as the ends of a string cannot be isolated, pulling quarks apart may require enough energy to produce the new quarks needed to form new nucleons. This may be the most suitable point to stop the discussion at the fundamental-particle end.

CHAPTER 2

water

Hydrides

2.1 Hydrogen Molecules

The simplest molecule must be the combination of two of the simplest atoms. Often it is the only molecule discussed in a quantum-mechanics class, leaving the impression that other molecules are just more complicated versions. The view here, in contrast, is that all other molecules are totally different. We begin with the simpler cases, the hydrogen compounds.

To understand the molecule H_2 we begin with the atom He, with its two electrons of energy $\varepsilon_s = -24.98$ eV from Table 1.1, and the electron density for each of the two electrons as approximately

$$\rho_0(r) = \frac{\mu^3}{\pi} e^{-2\mu r}, \tag{2.1}$$

from Eq. (1.3) with the $\mu = 2.56$ Å obtained from Eq. (1.2). We then do our *theoretical alchemy*, conceptually changing from one element to another by shifting protons among nuclei, as will be useful at many points in this book. We initially freeze the electron density to that in Eq. (2.1), though we soon find that a better approximation is to use a μ evaluated for the constituent atoms, here hydrogen. We separate the two protons in the nucleus (not worrying about neutrons) and let the two protons move to the position which minimizes the electrostatic energy. This is the hydrogen molecule, in the approximation that we neglect the change in the electron states due to the shift of the protons. [Such a picture is not new here, and is sometimes called the "united-atom model".] This calculation of electrostatic energy of two charges of $+e$ in a negative charge distribution of $-2e\rho_0(r)$ is completely straightforward, though a little intricate. It is carried out analytically in Appendix 2A, leading to the total energy if the two protons are at positions $\pm r$ from the center, along a diameter of the molecule. The result is

$$E(r) = 4e^2(\mu + \frac{1}{r})e^{-2\mu r} - \frac{7}{2}\frac{e^2}{r} + \frac{11}{8}e^2\mu + \frac{\hbar^2\mu^2}{m}. \tag{2.2}$$

The first two terms are the interaction of the two protons with the electronic charge plus the repulsion between the two protons $(e^2/2r)$. The third and fourth terms are the interaction energy between the two electrons and the kinetic energy of the two electrons and do not depend upon r.

The result is shown in Fig. 2.1 using $\mu = 2.56$ Å^{-1} appropriate to the neon charge density. We see that it has a minimum far from the observed

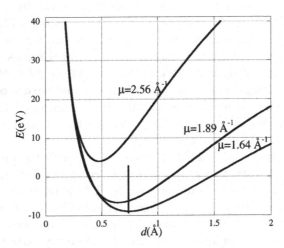

Fig. 2.1. The energy of two protons in the electron charge cloud as function of the distance between them d, $\mu = 2.56$ Å^{-1} for the helium distribution, 1.89 appropriate to hydrogen atomic states, and 1.64 chosen to fit the observed spacing in H, shown by the vertical line.

minimum of 0.74 Å indicated by the vertical line. It was clearly not appropriate to hold the charge cloud fixed as the protons moved. We might imagine that near one proton the charge cloud was more similar to that of hydrogen with the other proton and second electron having a small influence, suggesting a μ deduced from Eq. (1.2) using the hydrogen s-state energy of −13.61 eV from Table 1.1. Indeed that choice, corresponding to $\mu = 1.89$ Å^{-1}, brings the minimum much closer to the observed value, so we might proceed with that choice, and we do so in this case and make the corresponding choice for the central hydrides. Shifting it further to 1.64 Å^{-1} would be required to bring the spacing into agreement with experiment. We expect that the picture is qualitatively correct with the spacing dominated by the Coulomb repulsion between the two protons. This feature that the nuclear positions are determined by Coulomb interactions will only be true for hydrides.

It is possible to estimate the bonding energy of H$_2$ from this calculation by subtracting the energy of two isolated hydrogen atoms in the same approximation. For isolated atoms there is no electron-electron interaction (it is cancelled by the proton charge from the other atom), only the electron-proton interaction and the electron kinetic energy which leads to the electron-state energy of Table 1.1, the −13.61 eV of the hydrogen atom. Thus we are comparing our minimum in Fig. 2.1 with −27.2 eV and finding the energy higher in the molecule, rather than lower by the observed cohesion, −4.36 eV per molecule. This is a common difficulty with approximate theories, that if we compare energies for quite different situations, the small differences may be quite far off. Here the difficulty is that our electron-electron interaction (calculated as 37.4 eV using the $\mu = 1.89$ Å^{-1}) is large and may be inaccurate, and arises only for the molecule. Changing it to 12.5 eV would bring the energy into agreement with experiment. We may expect similar uncertainty in bonding energies for the central hybrids for the same reason.

We can however estimate the frequency at which the protons would vibrate, using the curvature at the minimum energy in Fig. 1.1, to obtain 3.7×10^{14} per second for the starting $\mu = 2.56$ Å^{-1}, dropping to 1.9×10^{14} per second for the adjusted $\mu = 1.64$ Å^{-1} much closer to the observed frequency of 1.3×10^{14} per second. There is no such large cancellation here as there was for the cohesion. We could also calculate the electric polarizability, which we shall do for the diatomic molecules in the next chapter, but not for the special case of the hydrides.

2.2 Central Hydrides

Of considerably more interest are the central hydrides, including methane, ammonia, water and hydrogen fluoride. We may try a similar theoretical alchemy for them. We begin with the neon atom, with six electrons in p states at -23.14 eV. There are also two s states, approximately twice as deep in energy. We again take the charge distributions to be given by Eq. (2.1), now with $\mu = 2.46$ Å^{-1} for the p states and 3.71 Å^{-1} for the s states. We look again at the comparison of this charge density with the full Hartree-Fock result, as in Fig. 1.1 which we repeat in Fig. 2.2. From both the full calculation and the approximate forms the charge distribution at large distances from the s states is quite a bit smaller than that from the p states. Generally, s states will extend further out from the atom than p states of the *same* energy [This is seen from classical-physics arguments in Appendix 2B.] but here with the s-state energy twice the p state energy, that energy difference wins. Further, there are six p electrons and only two s electrons, so we neglect the charge distribution from the s electrons and proceed with six times the $\rho_0(r)$ as for Eq. (2.1) with $\mu = 2.46$ Å^{-1} for Ne.

We begin by removing one proton from the neon nucleus, making it a fluorine nucleus and let that proton come to its position of minimum electrostatic energy, as we did for helium, but keeping the fluorine nucleus at the center. The changes in the calculation from that for helium are

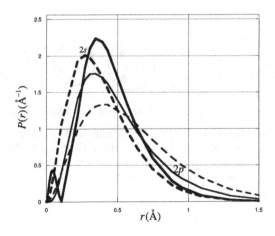

Fig. 2.2. The radial distribution of charge for neon, heavy lines for valence $2s$ states, light lines for $2p$ states, from Fig. 1.1, dashed from Eq. (1.4).

indicated in Appendix 2A. It leads to a form for the energy similar to that for H_2,

$$E(r) = 6\left(e^2(\mu + \frac{1}{r})e^{-2\mu r} - \frac{e^2}{r} \right) + 5\frac{e^2}{r} - 30e^2\mu + 15\frac{11}{8}e^2\mu + 3\frac{\hbar^2\mu^2}{m} \qquad (2.3)$$

and curves similar to that in Fig. 2.1. It is perhaps remarkable to still have simple algebraic forms. Using the $\mu = 2.46$ Å^{-1} for neon leads to a minimum, as in Fig. 2.1, at 0.93 Å, very close to the observed 0.92 Å. It would be raised to 1.00 Å with the $\mu = 2.28$ Å^{-1} ob2.1tained from the ε_p for fluorine. We find, in contrast to H_2, that we do better to neglect this approximation to the relaxation of the electron cloud, but that will not be true for the other central hydrides, as we shall see in Table 2.1. We regard these generally smaller μ's as indicating an expanded charge distribution so we have represented each distribution by a larger circle in Fig. 2.3. Our approach then will be to use the μ based upon the central-atom p state to estimate the molecular spacing, but then adjust the μ to obtain the observed spacing and a suitable model for calculating other properties.

When we treat water we shall allow the central nucleus to move from the center of the charge cloud, and discuss how we calculate the additional terms in the energy. The result will appear adequate also for ammonia. Using that approach for HF here we find that the fluorine nucleus moves off center by 0.055 Å.

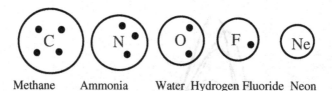

Methane   Ammonia   Water Hydrogen Fluoride Neon

Fig. 2.3. Representations of the central hybrids, with successive removal of a proton from the nucleus as we move from right to left. The "F" for example is the nucleus with nine protons and the dot, the other proton. The charge cloud is found to expand to the left, so the circles have been scaled up in proportion.

Table 2.1. Spacing in the central hydrides, that with μ estimated from the p-state energy of the central atom, and the μ needed to bring the spacing, and the molecular geometry, into accord with experiment.

	CH_4	NH_3	H_2O	HF
Observed d (Å)	1.10	1.01	0.96	0.92
d for μ from ε_p	0.77	0.81	0.88	1.00
μ (Å^{-1}) from ε_p	1.70	1.91	2.10	2.28
Needed μ (Å^{-1})	1.19	1.53	1.85	2.48

The estimate of the cohesive energy for HF with this formula is not successful, we presume for the same reason it was not for H_2, With the fluorine μ we obtain an energy of -207 eV, and for the separated fluorine and hydrogen atoms -284 eV, incorrectly a lower estimated energy for the isolated atoms than for the molecule. With the huge electron-electron interaction term, 737 eV in comparison to the observed cohesive energy of -5.89 eV for HF, it is not surprising. We abandon this approach for the cohesion of the central hydrides, and consider alternative approaches in the next section.

For water we pull a second proton from the nucleus and let the two protons adjust to the potential arising from the six-electron charge cloud. With oxygen at the center, we find the minimum energy (see Appendix 2A) if they lie on the same diameter, and with the neon charge density, at 0.75 Å from the center. With the μ appropriate to oxygen we obtain 0.88 Å from the center, not so far from the observed 0.96 Å, as indicated in Table 2.1. The fact that the expansion of the charge cloud is even greater than that suggested by the use of the smaller μ for oxygen rather than neon is a little surprising. Certainly removing a proton from the nucleus expands the charge cloud, and separate calculations suggest that adding it back to the system should expand it again somewhat. That seems not to be the case for water, ammonia and methane.

More importantly, in water the two protons move away for the molecular diameter, *only* because the electron charge density is not spherical around the oxygen nucleus, but shifts and deforms as the protons move from the positions on the diameter which we assumed. We should correspondingly allow the oxygen to move from the center of our distribution (representing a shift of the entire charge distribution relative to the nucleus), and at the same time allow the protons to seek their position of minimum energy. For this calculation we need the change in energy as the oxygen nucleus moves **b** from the center of the charge distribution. We discuss the estimate of this energy in Appendix 2A, where we find an energy of $\frac{9}{8}\hbar^2\mu^4b^2/m$ for small b, obtained from the calculated polarizability, and we added a quartic term proportional to $\hbar^2\mu^6b^4/m$. Indeed the total electrostatic energy dropped with the distortion, independent of that quartic term. We obtained the observed bend of $105°$ and oxygen-hydrogen spacings of 0.96 Å by adjusting the μ from 2.10 Å^{-1} to 1.85 Å^{-1} and choosing a coefficient of 203 for the quartic term. At the energy minimum the oxygen was displaced from the center by 0.065 Å. Considering our experience with HF, we did not try to estimate the cohesion of water.

A similar circumstance arises as we remove an additional proton to form a nitrogen nucleus. If that nucleus is at the center of the electron cloud, the three protons will form an equilateral triangle centered on the nucleus. Using the $\mu = 1.91$ Å^{-1}, obtained for the nitrogen p state, the minimum energy is found at 0.81 Å, compared to the observed 1.01 Å. We then allowed the nitrogen nucleus to move off center, and included the same terms and the same coefficients for the b^2 and b^4 terms as for water. The energy again dropped as the nitrogen nucleus and the protons moved, and a further shift of μ to 1.527 Å^{-1} was required to bring the spacing to the observed distance with $b = 0.08$ Å at the minimum energy, with an angle between protons of $109°$, close to the observed $107°$.

Finally, for methane, with one more proton removed from the nucleus and a μ for the carbon p state, the minimum energy comes with a tetrahedral arrangement of protons at a carbon-hydrogen spacing of 0.53 eV, compared to the observed 1.10 Å. Reducing μ to 1.18 Å^{-1} brings the spacing up to the observed value. Methane is the end of this central-hydride story. The next element, boron, forms as B_2H_6, rather than BH_5. Then BeH_2 and LiH are based upon the s, rather than the p, states, as were He and H_2.

We have found the simplest picture based upon the neon charge density, but a μ based upon the central atom ε_p, is consistent with the observed geometry in all cases, and that reasonable scalings of the decay parameter μ indicated in Table 2.1 bring the structures to their observed spacings and angles. We also note that in our picture the angles between the protons have nothing to do with a construction of tetrahedral hybrid orbitals as often assumed. Such hybrids will however be central to our understanding of the covalent solids and we shall revisit the hydrides after treating diatomic molecules.

2.3 Cohesion in the Central Hydrides

Our total energies based on the united-atom approach failed to give us reasonable estimates of the cohesion because the very large electron-electron interactions, calculated separately for the atoms and solid, were not adequate. To estimate cohesion for this case we require an outlook which does not separately estimate these large energies for the molecule and the atoms. The way we shall do this for ionic systems such as NaCl is to bring the neutral atoms together to the observed spacing, and assume that the Coulomb energy of the electrons does not change importantly as the electron is transferred from the Na to the Cl, so that the energy gain can be estimated as the corresponding $\varepsilon_p - \varepsilon_s$. Doing that for HF using Table 1.1 gives -6.26 eV in good accord with the experimental -5.89 eV. The same approach for water transfers two electrons, with an estimated cohesion of -6.32 eV, somewhat under the observed cohesion of -9.53 eV. Then for NH_3, with the p-state energy very nearly equal to hydrogen s-state energy, we would estimate a cohesion of -0.69 eV; the real cohesion of -8.19 eV cannot at all be described in such ionic terms. Going finally to methane, transferring four electron, in this case to the hydrogen rather than from it, suggests -10.16 eV in comparison to the observed -17.17 eV. The ionic limit again becomes at least meaningful, though not very accurate.

A similar situation to this series arises for HF, HCl, HBr, and HI. Our estimate for HF was fine, but in HCl the p-state energy is close to the hydrogen s-state energy and the ionic limit makes no sense. Beyond that again hydrogen receives the electrons and the estimates grow toward to observed values. In Chapter 3 we shall deal with these covalent contributions to the energy when the coupled states are of comparable energy and find a more acceptable description of the cohesion of both series of central hydrides.

We may note in passing that in Table 1.1 we placed hydrogen above fluorine in the table, which is appropriate for the heavier halogens and for methane where the hydrogen "receives" electrons in the bonding. Only for H_2O and HF would it be more appropriate to place it above lithium.

2.4 Ice and the Hydrogen Bond

Below 0°C water molecules form ice, which has many structures, all of which can be schematically represented as in Fig. 2.4 (a), and discussed in detail in Appendix 2C. The principal attraction between water molecules arises from the intricate Coulomb interactions between these systems of charges favoring each proton to be near the "back" side of the neighboring molecule as shown. Protons are sufficiently light that one may tunnel into the neighboring molecule as another cooperatively tunnels out of that molecule so that this *hydrogen bond* between molecules is not a simple classical situation as sketched in Fig. 2C(a), but similar to the *chemical bonds* we describe in Chapter 3, with the proton playing the role that the electrons play there, described in more detail in Appendix 2C. We can sketch the hydrogen bond between oxygen ions as the dumbbells shown in Fig. 2.4(b), each dumbbell representing a single proton. These are also the bonds which hold together the double strands of DNA. They are important there in that they are very weak in the sense that the strands are readily separated and recombined in cell division. However, they are strong enough here to bring the boiling point of water to its 100°C.

In the hexagonal structure of Fig. 2.4 each oxygen to the right of a horizontal hydrogen bond (call those sites R sites) might be moved out of the plane relative to those on the left of a horizontal hydrogen bond (L sites), forming a bilayer, two planes. (See, for example, Feibelman, 2010.) Each oxygen has three neighbors in the other plane of its bilayer, requiring 1½ protons per oxygen, leaving one proton per R site (or per L site). That proton could provide hydrogen bonding between each R site and the L sites in a neighboring bilayer above this one. The bonding to the bilayer below this would then be provided by the proton from that other bilayer. Every hydrogen bond is then identical, and there is no local electric dipole

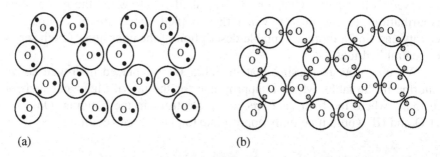

(a)   (b)

Fig. 2.4. A representation of the structure of ice, in part (a) with each proton near a neighboring molecule. As they tunnel between molecules they provide a hydrogen bond between the molecules, with each proton indicated by a dumbbell in part (b). In the real structure the hexagons are corrugated, or tessellated, forming a bilayer and these bilayers stacked upon each other with hydrogen bonds between the bilayers.

associated with that bond, and no dipole associated with each molecule, as there was for the isolated molecules. The energy difference between this hydrogen-bond state, and the corresponding antibonding state is very small, so that adding the two wavefunctions, which localizes the proton on one molecule, is also an *approximate* eigenstate. It is a state which might for example be appropriate at a surface. However, for understanding the structure of ice the dumbbell state is a better starting point, and a better way to think about the hydrogen bond.

2.5 Liquids and Solutions

Neon atoms are only weakly attracted to each other by van-der-Waals forces, which we described in Section 1.5, and repel each other at short distances as we shall describe in detail for general systems in Chapter 3. At low temperatures these forces result in a solid neon very much like marbles in a box. They have a structure (face-centered cubic, Fig. 4B.1) which has the densest possible packing for hard spheres. At $25°K$ it melts, and the volume increases by approximately the same percentage that the volume occupied by marbles packed in a box would increase if the box were shaken up. That is in fact a very reasonable way to think of the structure of the liquid inert gases, but in liquid neon the atoms are continually in motion within such an irregular structure. At $27°K$ it evaporates to the gas phase.

Above $0°C$ when ice melts, it forms in a structure much more like the neon liquid, which we picture as marbles randomly stacked in a box. In contrast to neon, the volume of water decreases on melting because of the open structure of ice before melting, indicated in Fig. 2.4. Again in the

Fig. 2.5. A schematic representation of liquid water, more closely packed than ice, but still with a strong correlation of the protons to energetically favorable positions relative to their neighbors.

liquid, the protons seek to orient toward the back side of neighboring ions, as in ice, and as in Fig. 2.5 for the liquid. In this case there are too few protons to make hydrogen bonds with each neighbor and in addition the molecules are also rotating as they undergo motion in the liquid. One can imagine it, and one could of course model it with a molecular-dynamics calculation, but it is a complex state of affairs.

We may also expect analogous structures for the solid and liquid states of the other central hydrides. Of more interest is a solution of one of these central hydrides in water. One could imagine, for example, replacing one or two of the water molecules in Fig. 2.5 by a hydrogen-fluoride molecule. It turns out that if one did, the energy would be lowered further if the proton from the hydrogen-fluoride molecule transferred to be a third proton in one of the water molecules, ultimately because it is then near an oxygen nucleus of smaller positive charge than for the fluorine nucleus it left. The first becomes a fluorine *ion* with a charge of $-e$ and the water with the extra proton is called a *hydronium ion*, with a charge of $+e$. The hydronium will take a structure like ammonia, with the three protons in a plane displaced from the oxygen nucleus and we may expect its charge cloud to shrink in comparison to water because of the extra proton; similarly the fluorine ion will expand. If these were isolated molecules, far from each other, this transfer would actually cost *extra* energy, but if we then brought them

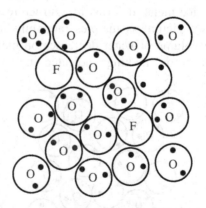

Fig. 2.6. A schematic representation of hydrofluoric acid. formed by the solution of an FH molecule in water, and proton transfers producing F^- ions and H_3O^+ ions. The F^- ion, which was smaller as an HF, expands and the H_3O^+ contracts.

together we would gain energy from the Coulomb attraction between the two charged ions, and when they are in the liquid their neighbors can gain energy by orienting their protons favorably relative to their neighbors. This is not difficult to imagine, and is illustrated in Fig. 2.6. It would be difficult to calculate in detail, but again it could be modeled by molecular dynamics. We will find a similar charge transfer that occurs in the condensed system which would not occur for isolated atoms when we form ionic crystals in the rock-salt structure. There it will involve electron transfer, rather than proton transfer.

We can similarly imagine substituting an ammonia molecule for a water molecule, and again energy is gained by a proton transfer, in this case from a water molecule to the ammonia. The resulting positively charged *ammonium ion* resembles methane, with four protons in a tetrahedron centered on the nucleus. It was initially larger than the water molecule but shrinks with the extra proton. The water molecule which *gave up* the proton forms a negatively charged *hydroxyl ion*, resembling the FH molecule, with a single proton, and expands in comparison to water. This ammonium hydroxide solution is illustrated in Fig. 2.7.

As we should expect, if we were to dissolve both hydrogen fluoride and ammonia in water, with both H_3O^+ ions and HO^- ions present, the energy will be lowered if we transfer the extra proton between them to produce two neutral H_2O molecules. This is simply the neutralization of the acid by the base. This results in the ammonium fluoride solution illustrated in Fig. 2.8.

Fig. 2.7. An ammonium hydroxide solution, containing NH_4^+ ions and OH^- ions.

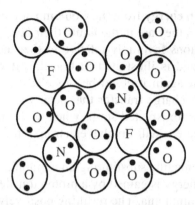

Fig. 2.8. An ammonium fluoride solution, obtained by mixing Figs. 2.6 and 2.7 and transferring protons between the H_3O^+ and HO^- ions to regain only H_2O.

We should use theoretical alchemy once more before moving to other molecular types. If for neon we *add* one proton we obtain a sodium nucleus and a positively charged sodium ion, unless we also add an electron. Here we have again changed the total nuclear charge in the neon charge cloud and it will shrink considerably. We could represent that charge cloud again using Eq. (2.1) but with a μ based upon the energy of the $2p$ core state of atomic sodium. We now substitute the resulting sodium ion for the NH_4^+ ammonium ion in Fig. 2.8 to obtain the sodium fluoride solution shown in Fig. 2.9.

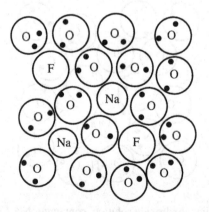

Fig. 2.9. A sodium fluoride solution. The sodium ion is drawn smaller because of the extra nuclear charge contained within the sphere.

Finally, we can imagine bringing some metallic sodium in contact with the HF of Fig. 2.6. The energy will be lowered if a sodium atom moves into the acid if it also places its electron at a site of lower energy, at the positively charged hydronium ion to make it neutral. Then with a second neutral hydronium, from another entering sodium atom, the energy will be lowered still further as a neutral hydrogen molecule emerges, leaving two neutral water molecules behind; hydrogen bubbles from the acid near the dissolving metal.

With these few cases we recognize that we have a conceptual model, and the basis of a model for calculation, for a vast array of systems, If we replace fluorine by chlorine, or other elements below it in the periodic table, the same description applies. For a calculation, only the p-state energy which determines the μ of Eq. (2.1) changes. The same is true for sodium and all of the other alkali metals. The same description will apply also for more complicated ions, such as the sulphate ion. Some such extensions may suggest molecules which are not in fact stable in nature, but if they are, the description should apply. One such problem is familiar if we move beyond the alkali metals to the divalent metals such as calcium. We might think that the corresponding Ca^{2+} ions could be incorporated into water just as were the Na^+ ions, but in fact they are largely insoluble in water, presumable because of their larger charge. Our simple conceptual model has not been applied to these questions.

2.6 Water near Metal Surfaces

One use for these representations of the central hydrides might be in the study of their behavior at a solid surface, such as a metal. This is rather straightforward, but we were surprised at the results. One major source of

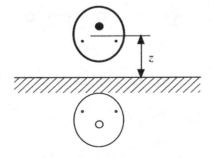

Fig. 2.10. A water molecule above a metal surface and its image charges.

interaction between the molecule and the metal is from the images of the charges in the molecule formed by the metal. This will in fact be the dominant interaction as the molecule moves far enough from the surface. Water is an interesting case and it is illustrated in Fig. 2.10.

We may first hold the oxygen nucleus and protons at there initial positions relative to the spherical electronic charge and calculate the electrostatic energy as a function of the water position z relative to the surface. For the position shown, with the protons closest to the surface, the energy varies as $-2.18/z^3$ eV-Å^3 at large z appropriate to a dipole interacting with its image, rises above that curve at smaller z but then drops and crosses it at 1.7 Å. If the oxygen is flipped over, with the oxygen nucleus closest to the surface the limit at large z is the same, but it remains lower in energy than the geometry of Fig. 2.10 down to 2.7 Å, at which z the energy rises above that of the Fig. 2.10 geometry. Thus, when it is close, the geometry shown is favored. If we then allow the protons and the nucleus to move within the electron cloud, as described in Appendix 2A, the difference is accentuated and at $z < 1.5$ Å the protons leave the charge cloud and enter the metal. This is indeed a possible behavior, but so are other scenarios. We may try to match up this picture with what is known about water at metal surfaces.

Little is known with any certainty concerning individual molecules at the metal surface, but Feibelman (2010) has recently given a careful account of the first, wetting, layers of water on solids, and we summarize some of that account, much of which dealt with water on transition metals such as ruthenium.

Ruthenium is a hexagonal-close-packed metal, with a surface which is represented in Fig. 2.11. The small circles are ruthenium atoms forming a

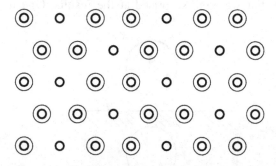

Fig. 2.11. Small circles are ruthenium atoms in a surface plane, perpendicular to the c axis of the crystal. The larger circles are oxygen ions in a bilayer in the ice structure as in Fig. 2.4.

triangular lattice. A bilayer of ice, which was shown in Fig. 2.4, is represented by the larger circles and superimposed on the ruthenium lattice in Fig. 2.11. The 2.60 Å distance between oxygen ions in Fig. 2.4 very nearly matches the 2.71 Å spacing in the ruthenium surface plane, so the oxygen bilayer has been stretched to fit. Feibelman notes that for decades this has been the accepted picture of the wetting layer of water on such a surface, a plausible picture. There would be hydrogen bonds between each oxygen ion in the picture and the extra proton on half of these ions was envisaged as on top of the upper ion plane, away from the metal, with the oxygen molecules in the lower plane of the bilayer making an electronic bond with the surface ruthenium atoms.

This would seem inconsistent with our finding that when the water is close to the surface, the protons would lie below the oxygen, and that when it came even closer, a proton entered the metal. However, Feibelman indicated that recent evidence indicates that the accepted picture is *not* correct. It was found that the oxygen layer is quite coplanar, suggesting that the extra proton goes to the remaining uncovered Ru in Figure 2.11. There are indeed exactly the number of surface sites needed. That is indeed consistent with our expectation of the proton entering the metal which we had replaced by a simple conductor. However, there was also experimental indication that at low temperatures a traditional bilayer of intact molecules can occur. It was suggested that excitation was required to reach the dissociated state with one proton gone from the molecule, an activation analogous to that we shall find necessary to form the dimerized reconstruction of the silicon (100) surface in Section 5.1.

The structure of that molecular film is not known, but knowledge of "rosettes" gives us a clue. These are clusters of seven hexagons of the kind shown in Fig. 2.11. There are 24 molecules in each rosette, and there is indication that the outer twelve have an extra proton, below the oxygen and over a ruthenium, consistent with our expectation of protons on the metallic side. Further, density-functional calculations also indicate protons either in the oxygen plane or between the oxygen and the surface. Feibelman emphasized the difficulty in having calculations sufficiently accurate to distinguish these different possibilities with very nearly equal energies, even with the best density-function-theory calculations. We are not seeking accuracy on any such scale, but our simple picture can help with interpretation. It is comforting for us that these recent results do not support the dangling-hydrogen picture which has been traditional, but inconsistent with our expectations. The positions of the protons in the wetting layer are important since without dangling hydrogen atoms above the oxygen, bonding of additional water molecules becomes difficult, leaving the wetting layer hydrophobic.

We should not forget that our simple picture replaced the metal by a conductor with no other structure, so the matching of the bilayer with the metal lattice constant would not come up in that description. Completing the picture will require going beyond the simple model of water as positive nuclear charges and protons moving in spherical electron clouds, which we shall do at the end of Chapter 3. It will also require discussing the structure of the metals as we shall do in Chapter 4, where we again discuss water molecules at the metal surface in Section 4.5. Finally, since ruthenium is a transition metal, we should include the effects of d states in the metal, which we do in general in Chapter 7, though not in the context of water on the surface.

air

CHAPTER 3

Molecules

We turn next to simple diatomic molecules, such as oxygen and nitrogen, held together by familiar two-electron covalent bonds. The top row of the periodic table, where oxygen and nitrogen appear, is the first row containing p states, and it seems that such first rows are always more complicated than the later rows, and we will find our approximations make larger errors in this first row than in later rows. Even in the simple Extended-Hückel Theory (Hoffmann, 1963) of covalent bonding, a scale factor 1.75 needed to be introduced, which turns out would have been essentially one if applied to heavier elements. An extra complexity similar to that for the $2p$ row will arise also in the $3d$ transition metals and the $4f$ rare-earth metals.

3.1 Molecular Orbitals

Molecules other than hydrides will generally require the construction of molecular orbitals, which contain orbitals from each of the constituent atoms. We consider the simple case of a molecular orbital based upon one valence state from each of two atoms, which might be s states in lithium atoms, and write the molecular orbital

$$|MO> = u_1|s_1> + u_2|s_2> . \qquad (3.1)$$

The expectation value $<E> = <MO|H|MO>/<MO|MO>$ of the Hamiltonian for this orbital is

$$<E> = \frac{u_1^2 <s_1|H|s_1> + 2u_1u_2 <s_1|H|s_2> + u_2^2 <s_2|H|s_2>}{u_1^2 + 2u_1u_2 <s_1|s_2> + u_2^2}, \tag{3.2}$$

where we took the atomic states to be real and normalized, $<s_1|s_1> = \int \psi_1(\mathbf{r})^2 d^3r = <s_2|s_2> = 1$, but not orthogonal to each other. $<s_1|s_2> \neq 0$. We obtain the best estimate of the energy of the molecular orbitals by a variational argument, minimizing this $<E>$ with respect to the u_1 and u_2. It was also a variational argument which Hartree (1928) used to formulate the one-electron approximation as discussed in Section 1.1. This provides the formal basis for most of the calculations in this book. It is so central to our approach that we carry it out for nonorthogonal orbitals in detail in Appendix 3A, but here we simplify by taking the orbitals orthogonal, and indicate what the additional effects of nonorthogonality are.

Thus we consider Eq. (3.2), but with $<s_1|s_2> = 0$. Then if the two atoms are the same, as for Li_2, the molecular orbitals can be taken to be even or odd, $u_2 = \pm u_1$ and $<s_1|H|s_1> = <s_2|H|s_2> = \varepsilon_s$, the atomic orbital energy of Table 1.1, we obtain immediately $<E> = \varepsilon_s \pm <s_1|H|s_2>$. The levels are split into bonding and antibonding orbitals. We call the magnitude of this coupling $<s_1|H|s_2>$ between the two atomic orbitals a *covalent energy*, and write it V_2. It is a coupling between *states*, not between electrons; the coupling between electrons would be e^2/r. For the specific case of coupling between s states $<s_1|H|s_2>$ is called $V_{ss\sigma}$, with the σ indicating zero angular momentum around the internuclear axis, and redundant for s states. If the energy of the two states were different, as for NaLi, we could specify their atomic energies as $\varepsilon_s^{\pm} = \pm V_3$ relative to the average, and V_3 is called the *polar energy*. The molecular-orbital energy is then readily found to be

$$<E> = \frac{\varepsilon_s^+ + \varepsilon_s^-}{2} \pm \sqrt{V_2^2 + V_3^2}, \tag{3.3}$$

reducing to $\varepsilon_s \pm V_2$ when the two energies are the same. We can obtain $V_3 = (\varepsilon_s^+ - \varepsilon_s^-)/2$ in each case from Table 1.1. V_2 could in principle be evaluated using the electron Hamiltonian and atomic wavefunctions, but we shall find another way.

It is also interesting to keep the $<s_1|s_2>$ to see its effect. We see in Appendix 3A that we can modify the definition of V_2 and V_3 and find that Eq. (3.3) is again obtained with an additional term $<s_1|s_2>V_2$. The effect of the nonorthogonality $<s_1|s_2>$ is to increase the energy of both states. It provides a repulsion between atoms arising from extra kinetic energy caused by overlapping the two electron clouds, and an essential ingredient in our understanding. We use (also in Appendix 3A) an assumption of Extended Hückel Theory (Hoffmann, 1963) that $<s_1|H|s_2>$ is proportional to the energy of the orbitals times $<s_1|s_2>$. We also introduce a scale factor λ to write the shift as $<s_1|s_2>V_2 = \lambda V_2^2/|\varepsilon|$, using the geometric mean of the two energies in the denominator if they are different, planning to adjust λ such that the calculated energy minimum comes at the observed spacing. Then Eq. (3.3) becomes

$$<E> = \frac{\varepsilon_s^+ + \varepsilon_s^-}{2} \pm \sqrt{V_2^2 + V_3^2} + \frac{\lambda V_2^2}{\sqrt{\varepsilon_s^+ \varepsilon_s^-}}. \tag{3.4}$$

Once we know V_2 and the observed spacing, we have everything we need to estimate the cohesion, and a wide variety of other properties.

3.2 Lithium Molecule

Thus for Li$_2$, with both atomic orbitals $\varepsilon_s = -5.34$ eV from Table 1.1, $V_3 = 0$. We shall shortly see that for s states a first estimate of the magnitude of the coupling is $V_2 = (\pi^2/8)\hbar^2/md^2$ in terms of the spacing d. The bond state in Eq. (3.4) is occupied with both spins to give an energy for the molecule, minus that for the two atoms, of

$$E_{Li2} = -2V_2 + 2\lambda V_2^2/|\varepsilon_s|, \tag{3.5}$$

minimum at $V_2 = |\varepsilon_s|/2\lambda$ with energy of $-V_2$. With the observed spacing of 2.67 Å, recalling that $\hbar^2/m = 7.62$ eV-Å^2 from Eq. (1.2), we find $V_2 = 1.32$ eV at that spacing and $\lambda = 2.02$. The energy $E_{Li2} = -1.32$ eV, the energy change in the formation of the molecule from free atoms, is to be compared with the observed -1.1 eV. We will not always do that well.

Having the electronic structure we can also estimate any other property of the molecule, which is not our goal here, but worth noting. We may estimate the vibrational frequency of the molecule since taking the second derivative of Eq. (3.5) with respect to d and evaluating it at the equilibrium spacing gives $\partial^2 E_{Li2} / \partial d^2 = 8V_2 / d^2$. We may let each atom move as $u \cos \omega t$ and equate the kinetic and potential energies to obtain

$$\omega_{Li}^2 = (2\pi\nu)^2 = \frac{16V_2}{M_{Li}d^2},$$ (3.6)

with M_{Li} the mass of the lithium nucleus for atomic weight 6.94. We obtain a frequency of $\nu = 10.2 \times 10^{12}$ cps. We might also estimate the polarization of the molecule for fields along the internuclear axis. We may do this just as we did for the atom in Eq. (1.6). The coupling between the bonding and antibonding states due to an electric field along the axis is $eEd/2$. It is squared, divided by the difference in energy $2|V_{ss\sigma}|$ between the two states, doubled for the two electrons and set equal to $\frac{1}{2}\alpha E^2$ to obtain a polarizability of

$$\alpha = \frac{e^2 d^2}{2|V_{ss\sigma}|},$$ (3.7)

equal to 38.7 Å^3. We did not find experimental values for either of these, but wanted to illustrate the possibilities, and these predictions are in the expected range.

3.3 Homopolar Molecules, Coupling

For Li_2 or NaLi we could consider the single s level on each and the algebra was very simple, with two levels the solution of a quadratic equation. As we make molecules from other atoms in the row, the $2p$ states must be included, and the molecular orbitals are linear combinations of four levels, requiring the solution of a fourth-order equation. Fortunately, for homopolar molecules we can use the symmetry to again reduce it to second-order equations. We sketch the procedure here, and give it in detail in Appendix 3B.

We first need the couplings between these additional orbitals, which we shall obtain in Section 4.2 by matching energy bands for simple metals obtained first with the extension of these molecular orbitals to three-

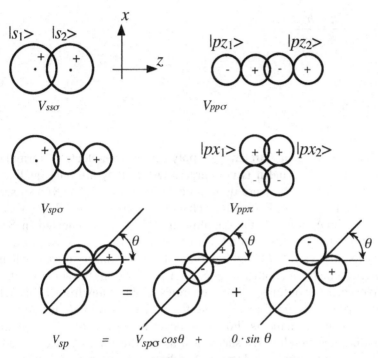

Fig. 3.1. Diagrams for the coupling between atomic orbitals on neighboring atoms, with values estimated in Eqs. (3.8) through (3.11). The lower diagram shows how couplings are calculated when the orbitals are not aligned with the internuclear axis.

dimensional metals, and second as free-electron-like bands. To define the couplings we choose a z axis along the axis of the molecule, and p states which are proportional to z/r are designated by $|pz\rangle$, etc., as illustrated in Fig. 3.1, as well as Fig. 1.1. Note again that the third subscript on the coupling is again the angular momentum around the internuclear axis. The final diagram shows how the couplings are calculated if this angular momentum is intermediate. From the matching in Section 4.2 we find

$$V_{ss\sigma} = -\frac{\pi^2}{8} \frac{\hbar^2}{md^2},$$

(3.8)

$$V_{pp\sigma} = \frac{3\pi^2}{8} \frac{\hbar^2}{md^2},$$

(3.9)

$$V_{pp\pi} = -\frac{\pi^2}{8}\frac{\hbar^2}{md^2},$$ (3.10)

and

$$V_{sp\sigma} = \frac{\pi}{2}\frac{\hbar^2}{md^2}.$$ (3.11)

It is a very long stretch to apply these free-electron formulae to diatomic molecules, and it is not surprising that they are only qualitatively correct here, though they are quite reliable in solids. Eq. (3.8) also seemed to do well for Li_2 earlier in this chapter. Again, $\hbar^2/m = 7.62$ eV-Å^2, so inserting d in Å gives us the coupling in eV. As we indicated in Section 1.2, writing states as linear combinations of atomic orbitals, as in Eq. (3.1), is traditionally called *LCAO* theory, but *tight-binding theory* when the couplings come from another source, such as Eqs. (3.8) through (3.10).

Given these couplings, and the energies of the states from Table 1.1, we have everything we need to calculate the molecular orbitals for any system made up of those elements. For homopolar molecules, we first obtain the antibonding and bonding $|s>$ states, just as we did for Li_2, as $\varepsilon_{sa,b} = \pm V_{ss\sigma} + \lambda V_{ss\sigma}^2/|\varepsilon_s|$, and similarly find the antibonding and bonding $|pz>$ states, taking our z direction along the molecular axis. We may add the coupling between the $|s>$ bonding state and the $|pz>$ bonding state, just as we did for Eq. (3.4) to obtain

$$\varepsilon_{MOb} = \frac{\varepsilon_{sb} + \varepsilon_{pb}}{2} \pm \sqrt{\left(\frac{\varepsilon_{pb} - \varepsilon_{sb}}{2}\right)^2 + V_{sp\sigma}^2 + \lambda V_{sp\sigma}^2/\sqrt{\varepsilon_s\varepsilon_p}}.$$ (3.12)

The minus sign will give the lowest-energy state of the molecule. There is also a pair of antisymmetric states with ε_{sb} replaced by ε_{sa} and ε_{pb} replaced by ε_{pa}. The antibonding and bonding π states, one pair based on $|px>$ states and the other on $|py>$ states, are immediate solutions for the molecules.

We can now proceed for the other elements in the top row of the periodic table, exactly as we did for Li_2, but keep track of the additional electrons and additional states, occupying the states of lowest energy. We simply add the energies of the occupied states, and subtract the energies which these electrons had in the free atom. We did this in Appendix 3B,

Table 3.1. Values of λ and scaling of the couplings of Eqs. (3.8)–(3.11) required to give the experimental values of the spacing and cohesion (both listed in Table 3B.1) for the diatomic molecules.

Molecule	λ	Scale factor for $V_{ll'm}$
Li_2	1.57	0.58
Be_2	0.830	0.5
B_2	1.015	0.261
C_2	1.038	0.252
N_2	1.207	0.245
O_2	1.650	0.205
F_2	3.32	0.114
Cl_2	1.51	0.346
Br_2	1.400	0.450
I_2	1.354	0.555

taking $\lambda = 1$, to obtain a guess for the spacing and the minimum total energies (an estimate of the cohesion) for *all* of the elementary diatomic molecules, the halogens as well as the first-row molecules. The results (Table 3B.1) were all of the right general magnitude, but not very impressive. Further, adjusting λ to obtain the correct spacing, as we did in the last section for Li_2, made the cohesive energies much too high. We attribute the discrepancies to the use of these free-electron couplings given above, and it seemed more to the point to adjust both a scale of all of the matrix elements, as well as λ, in order to obtain both the spacing and the cohesion correctly. The results of doing this appear in Table 3.1, where we give the factor by which all couplings should be scaled, and the λ to be used, to accomplish this. The differences of the scale factors from one give a measure of the error caused by the use of the free-electron formulae.

We can use the resulting scaling and λ for oxygen, for example, to obtain the energies as a function of spacing shown in Fig. 3.2. This is close to the fit which we used (Harrison, 2010) to treat oxygen interactions with surfaces of $La_{1-x}Sr_xMnO_3$. Thus it can be a simple and effective way for modeling molecules for such applications. As a test of our approach it shows the limitations to the accuracy of extending the free-electron formulae to systems such as the diatomic molecules. The required scalings seem to be largest for the right end of the first row of the Periodic Table. As in most things we discuss in this book, the behavior lower in the Table is simplest, and the top row the most anomalous. In all the diatomic molecules the curves are similar to those in Fig. 3.2 and the occupation of

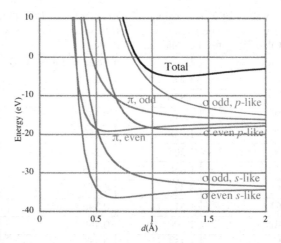

Fig. 3.2. Levels in the O_2 molecule, calculated with Eqs. (3.8) through (3.12), and the counterparts of Eq. (3.12) for the other states from Appendix 3B, with λ and the scaling of coupling from Table 3.1. The odd π states are half full, with two electrons, and all states at lower energy occupied (for $d > 0.7$ Å).

states was the same as in the ordering of levels for oxygen at the observed spacing shown in Fig. 3.2, for $d > 0.7$ Å.

In O_2 and C_2 we found the π states partly occupied. The state with both electrons of the same spin occupying the two states has lower energy due to quantum-mechanical exchange (Hund's Rule, discussed in Appendix 3C) so we find that the molecule has a net spin of one (½ plus ½). Thus O_2 is paramagnetic, with its magnetic moment aligning with applied magnetic fields. We expect the same for C_2 but we have not seen that discussed. We did not show those differences in Fig. 3.2.

It may be useful to relate these bonds and antibonds to traditional single, double, and triple bonds (*bond order*) before moving on to polar molecules and hydrocarbons. If we forced two neon atoms together, we could form molecular orbitals just as in Fig. 3.2, but with all states occupied the ± bonding energies cancel out and we are left with only the repulsive terms in λ. This is correct for neon, except for the independent van-der-Waals interaction which we described in Section 1.5. The F_2 molecular binding is traditionally described as a *single* two-electron σ bond, which we interpret here as having only a single antibonding molecular σ orbital empty for both spins, not canceling the lowering of the corresponding occupied orbitals. This provides the bonding of the molecule. In O_2 there are also two π antibonding orbitals (of the same spin) empty, making a π bond, a *double* bond for the molecule. In N_2, with the antibonding π orbitals of both spins empty, we have a *triple* bond. In C_2 we also empty one of the *bonding* π orbitals, so we reduce the bonding to a double bond, and in B_2 to a single bond. Be_2 by these standards has no bonding (and it does not form as a diatomic molecule), and Li_2 has again a single σ bond. At the beginning of this section we included only the s states in the calculation, neglecting the $V_{sp\sigma}$ in Eq. (3.12). We regard both treatments of Li_2 as meaningful, but with different orbitals included in the basis set by which we seek to describe the system, we need different scalings to make it agree with experimental parameters. There are many cases in this book in which we use alternative basis sets, taking both as meaningful, and sometimes learning what the parameters must be if both representations are to be meaningful. That is how we obtained the couplings given in Eqs. (3.8) through (3.11) and illustrated in Fig. 3.1.

3.4 Polar Molecules, Hybrids

We should note that the generalization of these molecular orbitals to molecules consisting of two types of atoms is completely straightforward using the atomic-state energies from Table 1.1 and the couplings from Eqs. (3.8) through (3.11). The difficulty is that we must now include for each σ state all four orbitals independently, rather than two even combinations or two odd combinations. The problem becomes the solution of a four-by-four, rather than a two-by two matrix, and therefore a numerical solution rather than the solution of quadratic equations. Thus we can proceed directly, and numerically, with molecules such as boron nitride (BN), cyanide (CN), or nitric oxide (NO), the latter two having odd numbers of electrons and therefore paramagnetic as was O_2.

We pick CN as a representative case of a polar diatomic molecule. In Appendix 3D we proceed with our parameters just as we did for O_2. We then solve the four simultaneous equations numerically using a scale factor of 0.249 and a $\lambda = 1.12$, obtained from averaging those for C and N from Table 3.1. We find curves much like those of Fig. 3.2, shown as Fig. 3.3(a). We add the energies of the nine electrons in the lowest states to obtain a total energy with a minimum of -6.73 eV at a spacing of 1.10 Å, compared to the observed -8.0 eV and 1.48 Å. We can also adjust λ to 0.97 and the scale factor to 0.52 to obtain the observed equilibrium spacing and energy.

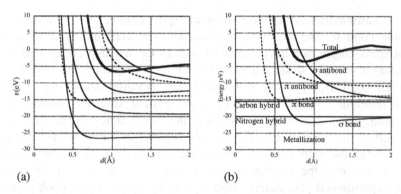

(a)   (b)

Fig. 3.3. The energy levels of CN, in part (a) based upon the full tight-binding calculation, the counterpart of Fig. 3.2, which showed the levels for O_2. The results in part (b) are based upon hybrid states.

There is a way to avoid this numerical procedure using hybrids, which will be particularly convenient in covalent solids. It also gives a clearer picture of the bonding than obtained from the numerical solution. Rather than expanding our molecular orbitals in the s and p states, we could equivalently use hybrid states on each atom for the σ states. As in Eq. (3.1), here on a single atom, we form the two hybrid states

$$|h^{\pm}> = (|s> \pm |pz>)/\sqrt{2} \qquad (3.13)$$

if it is part of a diatomic molecule oriented along the z axis. Each has an average energy

$$\varepsilon_h = <h^{\pm}|H|h^{\pm}> = (\varepsilon_s + \varepsilon_p)/2 \qquad (3.14)$$

based on the s- and p-state energies for that atom, but the hybrid is not an eigenstate with a single energy. We reserve the term *hybrids* for these combinations of states on the same atom. It is common practice to use it also when bonds are formed, to say that an electron on one atom hybridizes with the states on another atom but we feel that this is using the same term for a fundamentally different concept.

If we proceeded to solve for the molecular orbitals exactly, as we did above for the four atomic states entering the σ orbitals, we would obtain exactly the same molecular orbitals and energies. However, these hybrids offer an opportunity for a simplifying approximation. For the + sign in Eq.

(3.13) we may see from Fig. 3.1 that the positive wavefunctions on the right of the atom add, but cancel to the left so that the state *leans* to the right. Correspondingly, there is a strong coupling between a right-leaning hybrid on the left atom and a left-leaning hybrid on the right, with magnitude

$$V_2 = (V_{pp\sigma} + 2V_{sp\sigma} - V_{ss\sigma})/2 = -(2 + 4/\pi)V_{ss\sigma} = 4.04\hbar^2/md^2 . \tag{3.15}$$

All the contributions add with the same sign. On the other hand, a right-leaning hybrid is coupled to the other right-leaning hybrid by only magnitude $(V_{pp\sigma} + V_{ss\sigma})/2 = \hbar^2/md^2$ and two outward-leaning hybrids by a coupling of magnitude $(V_{pp\sigma} - 2V_{sp\sigma} - V_{ss\sigma})/2 = 0.73\hbar^2/md^2$. Thus it becomes a very reasonable approach to include only the coupling V_2 of Eq. (3.15) for the four σ molecular orbitals. We are left with two *nonbonding* states, outward-leaning hybrids sometimes called *dangling hybrids* (often less appropriately called dangling bonds), one for each atom, and bonding and antibonding states obtained exactly as for Eq. (3.4) as

$$\varepsilon_{MO}(\sigma, a_{or}b) = \tfrac{1}{2}(\varepsilon_h^+ + \varepsilon_h^-) \pm \sqrt{V_2^2 + V_3^2} + \lambda V_2^2/\sqrt{\varepsilon_h^+ \varepsilon_h^-} , \tag{3.16}$$

with the plus superscript now designating the hybrid with higher energy and the minus superscript designating the hybrid with lower energy. The a or b goes with the plus or minus, respectively, in front of the square root. The polar energy is given by $V_3 = (\varepsilon_h^+ - \varepsilon_h^-)/2$.

This reduction to only two coupled states is a considerable simplification in the case of polar systems where a numerical solution is otherwise required for expansion in four states. The π states are already of this form as

$$\varepsilon_{MO}(\pi, a_{or}b) = \tfrac{1}{2}(\varepsilon_p^+ + \varepsilon_p^-) \pm \sqrt{V_{pp\pi}^2 + V_3^2} + \lambda V_{pp\pi}^2/\sqrt{\varepsilon_p^+ \varepsilon_p^-} , \tag{3.17}$$

with $V_3 = (\varepsilon_p^+ - \varepsilon_p^-)/2$ for the π states and the plus before the square root again the a, or antibonding, state and the minus for the b, or bonding, state.

We applied this method to the CN molecule, described in detail in Appendix 3D, to obtain the energy levels shown in Fig. 3.3(b). The minimum energy of -3.66 eV comes at $d = 0.92$ Å, to be compared with the full calculation which gave -6.73 eV at $d = 1.10$ Å, closer to the observed -8.0 eV at 1.48 Å.

It is seen in Appendix 3D that the principal error in using hybrids did not come from the coupling between outward-leaning hybrids and the states on the other atom, which we neglected, but from an intra-atomic coupling between the outward-leaning hybrid and the inward-leaning hybrid *on the same atom*. This is readily seen using Eq. (3.13) to be

$$<h^{\pm}|H|h^{\pm}> = (\varepsilon_s^{\pm} - \varepsilon_p^{\pm})/2 \equiv -V_1 , \qquad (3.18)$$

with the magnitude V_1 called the *metallic* energy. Fortunately this effect, which in semiconductors is called *metallization* of the bond, can be corrected for at the end. We discuss how that is done in Appendix 3D. It is rather simple for a homopolar molecule, like O_2 where we could also have used hybrids, and is then found to give a metallization energy of

$$\varepsilon_{metal.} = V_2 - \sqrt{V_2^2 + 4V_1^2} , \qquad (3.19)$$

with the V_2 of Eq. (3.15). In semiconductors the metallic energy V_1 is ordinarily small enough compared to the covalent energy V_2 that we can use perturbation theory, or an expansion of Eq. (3.19) as $\varepsilon_{metal.} \approx -2V_1^2/V_2$, but for oxygen V_1 is much larger than V_2 and Eq. (3.19) is appropriate. The use of hybrids for oxygen is very inaccurate, but including metallization is seen in Appendix 3D to recover most of the error.

We also made application of metallization to CN in Appendix 3D. It was more complicated than the homopolar case, but we found as in O_2 that it removed most of the error arising from simplifying the problem with the use of hybrids. Again we could adjust our scaling and λ to bring it into agreement with experiment.

In the case of these polar molecules, the inclusion of metallization is so intricate that it would appear preferable to simply do the numerical solution of the four-by-four problem. In semiconductors the use of hybrids makes the difference between calculations of the properties with simple equations as opposed to major numerical density-functional calculations, so the benefits are much greater. Also in semiconductors the V_1 are small enough to allow perturbation theory, simplifying the calculation. Even in molecules, the use of hybrids makes the details of the bonding much more understandable. We separate the role of the dangling-hybrid states from that of the σ and π bonds, and the distinct contribution of metallization of the bonds, even if calculating it is so burdensome that we do better to return to the full numerical calculation of the four-by-four matrix.

We might well be concerned that for these *polar* molecules there is some charge density transferred to the more electronegative atom, that with deeper atomic levels, and that the Coulomb potentials arising from the transfer might raise the energies on the more electronegative side, reducing the V_3 obtained from neutral atoms. That is true, but we shall see in Chapter 6 that even for extreme cases such as crystalline rock salt, the Coulomb lowering of the states on the more electronegative atom by the positively charged neighbors approximately cancels the intra-atomic shift. A self-consistent calculation of these shifts would not be unreasonable but, on the scale of accuracy we are working with, it is not necessary.

We will at times need the coefficients in the molecular orbitals, as well as the energies of the states. We in fact needed them for the calculation of metallization of the bonds in CN in Appendix 3D. Obtaining the coefficients is simply a matter solving a pair of simultaneous linear algebraic equations, but it is easier to check the answer than to derive it. We first define a polarity α_p for the molecular orbital with energy given in Eq. (3.16) as

$$\alpha_p = \frac{V_3}{\sqrt{V_2^2 + V_3^2}}. \tag{3.20}$$

It is analogous to Pauling's ionicity (though his ionicity is closer to α_p^2) in that it is zero for a symmetric bond, and goes to one for an extremely polar bond. [The motivation for *ionicity* is however quite different. It is used to interpolate experimental quantities among compounds. α_p is a parameter entering theoretical estimates.] The coefficients of the states, written as in Eq. (3.1), are

$$u_1^b = \sqrt{\frac{1+\alpha_p}{2}}$$

and $\hspace{8cm}$ (3.21)

$$u_2^b = \sqrt{\frac{1-\alpha_p}{2}}$$

for the bonding state (the $-$ in Eq. (3.16) or (3.17)) if $V_3 = (\varepsilon_2 - \varepsilon_1)/2$. The sign of V_3 depends upon which state is more electronegative. For the antibonding state they are

$$u_1^a = \sqrt{\frac{1-\alpha_p}{2}}$$

and   (3.22)

$$u_2^a = -\sqrt{\frac{1+\alpha_p}{2}}.$$

Then the probability of an electron in the bond being found on the first site is $u_1^{b\,2}$ and the probability on the second site $u_2^{b\,2}$. This could be used to calculate the polarizability of the bond by shifting the two energies with an electric field E along the bond by $\pm eEd/2$, but we calculated polarizabilities before using the energy shift.

3.5 Electron-Electron Interactions

In these polar molecules the distinction between energy required to remove an electron from an atom (the ionization energies of Table 1.1) and the energy gain in adding an electron to a neutral atom (electron affinity, with an added U from Table 1.1) comes up again. We noted in Section 1.5 that the U should be reduced by the attraction, $-e^2/r$, by the effect of the positive charge left on the atom a distance r away, from which the electron came. In most of these diatomic molecules the two contributions very nearly cancelled, and we did not need to consider them. However, this will not always be true, and if we were to pull these diatomic molecules apart in a reaction, this U comes into play and we should consider its role. It will provide us with a means of including its effects when it is important.

This can easily be done in the context of the Li_2 molecule which we discussed in Section 3.2. We now think of this as a *two-electron* problem, involving the single s state on each atom. We again let the two electrons have opposite spin, but there are now four two-electron states: |1> with both electrons on the first atom, |2> with both electrons on the second atom, |3> with the up-spin electron on the first atom and the down-spin on the second, and |4> with the two electrons interchanged. There are no other two-electron states with opposite spins, only combinations of these. The energies of |1> and |2> are higher by this effective Coulomb interaction $U^* = U - e^2/r$, which we no longer ignore. Also there is the coupling $V_{ss\sigma}$ between the state |1> and the state |4> as well as with the state |3>, and similarly between the state |2> and both states |3> and |4>. We proceed variationally with the two-electron problem exactly as for the one-electron problem in Eq. (3.2), but now leading to four simultaneous equations in the four u_j. For this symmetric problem, the states are all either even or odd ($u_2 = \pm u_1$, and $u_4 = \pm u_3$) and it reduces to quadratic equations with solutions

$$E_{2el} = 2\varepsilon_s + \frac{U^*}{2} \pm \sqrt{\left(\frac{U^*}{2}\right)^2 + 4V_{ss\sigma}^2}. \qquad (3.23)$$

The lowest-energy state, the ground state of the molecule, comes with the minus sign. Note that if we take $U^* = 0$, we obtain $2\varepsilon_s - 2|V_{ss\sigma}|$, both electrons in a bonding state as we found in Section 3.2. If U^* is very large, we obtain $2\varepsilon_s$ with the coupling not important, but very significantly with no U^* contribution; the electrons are segregated to separate atoms. Had we proceeded without U^*, we would have formed bonds corresponding to a 50% probability of both electrons being on the same atom, and this would have remained true no matter how small $V_{ss\sigma}$ became. The difference is important new physics arising from the electron-electron interaction and why this calculation was given in Harrison (1999), 594ff. This was not important in this chapter where we formed molecules and the U^* was small, but we will need this correction for U^* when we consider molecules approaching a surface from a distance, and when we treat transition metals in Chapter 7.

3.6 Hydrocarbons

Our forming of the central hydrides by the successive removal of protons suggests how we might proceed with hydrocarbons using theoretical alchemy. We begin with the F_2 molecule, with a single bond as described above. We freeze the charge distribution and pull three protons out of each nucleus, producing carbon nuclei, and let them fall back on the two sides. The minimum electrostatic energy will put them approximately on three corners of a tetrahedron centered on the carbon atom, with the fourth tetrahedral corner aimed at the other carbon, as illustrated in Fig. 3.3. The two tetrahedra will be rotationally oriented with respect to each other such that the protons are not directly opposite each other, so the electrostatic energy is lower. We then let the electron cloud relax, which we saw for the central hydrides is a significant relaxation, but does not change the picture qualitatively. We have thus constructed ethane, C_2H_6, with a single bond between the carbon atoms.

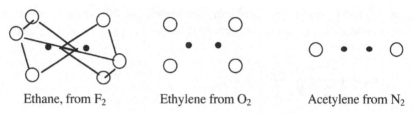

Ethane, from F_2 Ethylene from O_2 Acetylene from N_2

Fig. 3.3. Construction of organic molecules from the diatomic molecules by removing protons from the nuclei and letting them fall back into the electron cloud. All atoms in ethylene and acetylene are in the plane of the figure.

We next start with O_2 and repeat the process, removing two protons from each O nucleus to produce C nuclei, and let them fall back. In this case the minimum electrostatic energy would be *not* as shown in Fig. 3.3, but with the two protons on the right rotated $90°$ around the axis of the ethylene molecule. However the observed structure is that shown, with all four protons in the same plane. The cause for this is the presence of only two electrons in the antibonding π states of O_2. We have the option of placing them *both* in a py state, (with the axis vertical in Fig. 3.3) which would concentrate the electronic charge density in the plane of the figure, and favor the protons being arranged as shown; the electron density is not cylindrically symmetric. That is what happens. Actually, putting them both in a py state requires that they have different spin orientations (Pauli's principle), so this costs us the exchange energy which caused us in O_2 to put one electron in a px and one in a py state, making the charge density in O_2 cylindrically symmetric, and the molecule paramagnetic. However, the electrostatic energy wins over exchange in ethylene, and the molecule is not paramagnetic. If we could turn up the exchange, at some point exchange would win; the protons would rotate out of the plane and the molecule would become paramagnetic, but that does not happen in nature. The ethylene molecule, like the O_2 molecule has a double bond, including the px π bond, the contribution of which is not cancelled because the px π antibond is empty.

Finally we take N_2, remove one proton from each nucleus and let it fall back, to produce acetylene, as shown in Fig. 3.3, with the orientation of lowest electrostatic energy, cylindrically symmetric with both antibonding π states empty, and so a triple bond, like N_2.

This picture of the organic molecules extends beyond these two-carbon cases. Sulphur, selenium, and tellurium are in the same column with oxygen, and have the same valence electronic structure as oxygen but with different atomic-state energies. Their atoms form in long chains, as we shall see in Section 5.4, though oxygen happens only to occur as diatomic molecules because of its different state energies and much smaller spacing. We can nevertheless imagine oxygen as such a chain, with perpendicular σ bonds to its two nearest neighbors at 90° orientations. We remove two protons from each nucleus, to produce a carbon, and they fall back so that each carbon nucleus has four neighbors: two carbons and two hydrogens, giving us now *singly* bonded propane, butane, etc., depending on the length of the chain. [We would need actually to terminate the oxygen chains with fluorine, and remove three protons from those atoms to construct these molecules, but it still works.] For benzene we can imagine forming a hexagon with six nitrogen atoms and removing one proton from each nucleus. The bonding in the nitrogen ring is however tricky, resonant bonds which we shall deal with separately in Section 5.5.

3.7 Revisiting Hydrides

The bonding in these organic systems, as well as in the hydrides, is more usually discussed with the hydrogen atoms entering as full partners, with their own 1s orbitals entering the expansion of the molecular orbitals. It is different, but neither approach is necessarily wrong and we may proceed with the approach we used for diatomic molecules. For the hydrogen molecule we would introduce a coupling $V_{ss\sigma} = -(\pi^2/8)\hbar^2/md^2$ from Eq. (3.8) and a repulsion $\lambda V_{ss\sigma}^2/|\varepsilon_s|$ with two electrons in the molecular orbitals. An estimate of the spacing with $\lambda = 1$ yields $d = 0.53$ Å (actually exactly the Bohr radius because ε_s is one Rydberg, $e^4m/2\hbar^2$) in comparison to the observed 0.74 Å. Then we may scale the coupling and adjust λ to fit this observed d and the observed cohesion of -4.36 eV. This scales the coupling $V_{ss\sigma}$ by 0.268 and sets $\lambda = 1.56$, not out of line with what was required for the other diatomic molecules listed in Table 3.1.

For the central hydrides the dominant term in the coupling will be between the hydrogen s state at $\varepsilon_s = -13.61$ eV and the central-atom p state, and their coupling is estimated as $V_2 = V_{sp\sigma} = (\pi/2)\hbar^2/md^2$. For HF with $\varepsilon_p = -19.87$ eV we obtain $V_3 = 3.13$ eV and energy relative to separated atoms of

$$E = -2\sqrt{V_2^2 + V_3^2} + \frac{2\lambda V_2^2}{\sqrt{\varepsilon_s \varepsilon_p}}. \tag{3.24}$$

Note that without coupling, $V_2 = 0$, we obtain $-2V_3$, with both electrons on the fluorine, rather than one on each atom because we did not include the electron-electron interactions discussed in Section 3.5. Keeping V_2, λ can be adjusted so that the minimum energy is at the equilibrium spacing, giving

$$\lambda = \sqrt{\varepsilon_s \varepsilon_p} \Big/ 2\sqrt{V_2^2 + V_3^2}, \qquad (3.25)$$

and then the energy becomes

$$E = -2\sqrt{V_2^2 + V_3^2} + V_2^2 \Big/ \sqrt{V_2^2 + V_3^2}. \qquad (3.26)$$

This gives more cohesion than the observed -5.9 eV for all scaling of the coupling because it is always below the energy $-2V_3 = -6.26$ eV for the electron transfer with *no* coupling. However, with a small scale factor as in many cases in Table 3.1, Eq. (3.26) will be close to -6.26 eV and therefore not far from the experimental -5.9 eV and λ will approach the 2.63 obtained from Eq. (3.26) with $V_2 = 0$.

For H_2O we may again allow the protons to lie off the diameter by an angle θ, as in Appendix 2A, and we find two bonding states, one with a $V_2 = \sqrt{2}V_{sp\sigma}\cos\theta$ (for the oxygen pz state and the combination of hydrogen s states entering as $(|s1\rangle + |s2\rangle)/\sqrt{2}$ if the z axis is vertical in Fig. 2A.1). The other bonding state has a $V_2 = \sqrt{2}V_{sp\sigma}\sin\theta$ (with the px oxygen state and $(|s1\rangle - |s2\rangle)/\sqrt{2}$ entering). The energy is minimum with $\theta = \pi/2$, both bonds with the same energy and each doubly occupied, giving

$$E = -4\sqrt{V_2^2 + V_3^2} + 2V_2^2 \Big/ \sqrt{V_2^2 + V_3^2}, \qquad (3.27)$$

just twice Eq. (3.26). In hindsight this is clear; we could initially place the protons $90°$ from each other and form independent σ bonds with each, exactly like the HF bond. [These states are discussed in more detail in Appendix 4C.] There is also an argument based upon moments (Harrison

Table 3.2. Observed interatomic spacing and binding energy (the heat of formation of the molecule minus that of its constituents from Weast (1975).

	Observed d(Å)	Obs. E(eV)	Scale $V_{sp\sigma}$	λ
HF	0.92	−5.9	Small	2.17
HCl	1.27	−4.5	0.61	1.52
HBr	1.41	−3.8	0.61	1.76
HI	1.61	−3.1	0.42	2.59
H_2O	0.958	−9.6	0.300	1.80
H_2S	1.346	−7.6	0.51	1.79
H_2Se	1.47	−6.0	0.2	3.28
NH_3	0.918	−12.2	0.28	1.72
PH_3	1.41	−8.9	Small	2.8
AsH_3	1.519	−8.1	Small	2.59
CH_4	1.10	−17.3	0.324	1.57
SiH_4	1.48	−14.4	Small	2.69

1999, 94ff) that it is energetically favorable to form independent bonds, with no coupling between them, in this case a 90° orientation. In any case, Eq. (3.25) and a doubled Eq. (3.26) apply. Use of the free-electron $V_{sp\sigma}$ overestimates the cohesion by a factor of three, but we *can* choose a smaller value of $V_{sp\sigma} = 3.90$ eV, scaled down by a factor 0.300, which yields the observed cohesion of $E = -9.6$ eV, and then if λ is equal to 1.80 from Eq. (3.25) we have fit the observed spacing. Of course we are left with an angle between oxygen-hydrogen bonds of 90° rather than the observed 104.5° and a different concept for its origin than described in Chapter 2 and Appendix 2B. The larger angle is often attributed to Coulomb repulsion between the protons which was not included in this bonding description. We proceed to the other central hydrides and have collected the fitting parameters with the observed spacing and cohesion in Table 3.2. We included all the central hydrides for which we could find data. For NH_3 Eqs. (3.25) applies and (3.26) is tripled, just as it was doubled for water, with the three hydrogens at 90° from each other. For methane, each p state on the carbon is coupled to a combination of the four hydrogen s states of the form $(|s_1> + |s_2> - |s_3> - |s_4>)/2$, with different signs for the three p states. The coupling becomes $V_2 = 2V_{sp\sigma}/\sqrt{3}$ for each p state, and we may redo the counterparts of Eqs. (3.25) and (3.26). For the molecules for which, as in HF, we could not fit both d and energy, but were close with a small scale, we have indicated "small" in the table.

This view of the central hydrides is quite different from what we gave in Chapter 2, but the charge distributions are not so terribly different. We may construct them from the molecular orbitals of Eq. (3.1) using the coefficients Eq. (3.21) and the atomic wavefunctions of the form of Eq. (1.1). We needed also angular factors such as a factor x/r for the x-oriented p state. Adding the resulting charge density from the three p-like molecular orbitals for the water molecule gives the electron density shown in Fig. 3.4, plotted along four directions from the oxygen nucleus. The most interesting comparison is for the [−100] and [100] curves, relative to the [010] curve. In the midrange ($\rho(r) \approx 0.5$) the first two are displaced by −0.07 and +0.07 Å relative to [010]. With the simple shift of a spherical density that we assumed in Chapter 2 (for $b = 0.065$), these would have been displaced by −0.065 and +0.065 Å, respectively, very close to the same. The striking difference, of course, is that in the [110] direction, through one of the protons, there is the additional peak from the localization of charge around that proton. How important this is will of course depend upon just what property we are considering.

It will be important when we return to the problem of a water molecule coming to a metal surface. However, to do this in the context of molecular orbitals we must consider the metal in these terms, a more general application of the Linear Combination of Atomic Orbitals which we described here for molecules. We proceed to that next.

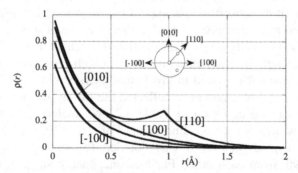

Fig. 3.4. The electron density for a water molecule, along four different directions, based upon molecular orbitals with our approximation (Eq. (1.1)) to the atomic orbitals. The protons are in the [110] and [1−10] directions from the oxygen nucleus.

tin

CHAPTER 4

Simple Metals

4.1 A Linear Chain

We shall begin with lithium, as we did in the preceding chapter, and review briefly the formation of energy bands. There is just one important atomic orbital and one electron per atom, but rather than two atoms, we consider a chain of atoms, spaced by $d = 3.03$ Å, the nearest-neighbor distance in metallic lithium. We can initially simplify it even further by bending it into a circle, so that we eliminate the ends of the chain, also called *periodic boundary conditions*. We again write the states as a linear combination of these atomic states, ψ_j on the jth atom, numbering as shown in Fig. 4.1(a),

$$\psi = \sum_j u_j \psi_j \qquad (4.1)$$

as in Eq. (3.1). We may evaluate the energy of the state, as in Eq. (3.2) for two atoms, assuming orthogonality of neighboring states, $<\psi_i|\psi_j> = \delta_{ij}$, with every $<\psi_i|H|\psi_i> = \varepsilon_s$ the same, and $<\psi_i|H|\psi_{i\pm1}> = V_{ss\sigma}$. Seeking an eigenstate is essentially evaluating the expectation value of that energy, $<\psi|H|\psi>/<\psi|\psi>$ and doing a variational calculation as in Appendix 3A, finding the minimum energy with respect to each of the coefficients, u_j. The variational conditions (e.g., Harrison, 2000, p. 88) become

$$\varepsilon_s u_j + V_{ss\sigma}(u_{j-1} + u_{j+1}) = \varepsilon u_j, \qquad (4.2)$$

For $j = 1, 2, 3, \ldots, N$. Floquet's Theorem, more usually called Bloch's Theorem, states that if every atom is identical, the coefficients for the atomic states can be written $u_j = e^{ikx}{}_j / \sqrt{N}$, where x_j is the position of the ion measured by x along the chain from the Nth atom, as indicated in Fig. 4.1(a). k characterizes each particular state and can take any value such that kNd is an integral multiple of 2π (so that it comes out even around the chain) and the $1/\sqrt{N}$ normalizes the state. This follows from the fact that for any eigenstate, if we move each coefficient the same number of steps d along the chain, the new state formed must also be an eigenstate, with the same energy. With the coefficients we have chosen it is the same state, multiplied by a phase factor e^{-ikT}, with T the distance the state was moved. These eigenstates may be thought of as waves, propagating with wavenumber k around the ring. We may confirm that they are solutions by substituting Eq. (4.1) into Eq. (4.2). Doing so leads to

$$\varepsilon_k = \varepsilon_s + 2V_{ss\sigma}\cos(kd). \qquad (4.3)$$

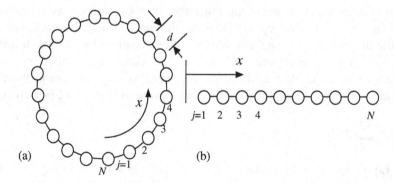

Fig. 4.1. A chain of N lithium atoms. In (a), they are bent into a ring; in (b), in a straight line.

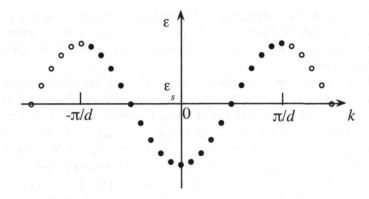

Fig. 4.2. The energy bands for the ring shown in Fig. 4.1(a). Each dot represents a state; those solid are each different and the empty dots are each equivalent to some solid dot.

This equation gives the *energy band* for this chain of atoms, joined at the ends, shown in Fig. 4.2. There are exactly as many different states as there are atoms in the chain (wavenumbers differing by $2\pi N d$ give the same coefficients u_i). The wavenumber range from $-\pi/d$ to $+\pi/d$ is called the *Brillouin Zone*, and contains all the different states. If there were twice as many atoms there would be twice as many dots, but they would lie on exactly the same line in the same Brillouin Zone. Two electrons can occupy any one state, so the ground state of the chain of lithium atoms will have all the states with energy below ε_s doubly occupied (and that at ε_s half occupied). As in the diatomic molecule we gain energy by occupying states lower in energy than ε_s.

These bands also describe completely the dynamics of electrons occupying the bands. By making wave packets we can easily obtain (e.g., Harrison (2000), 189ff) expressions for the velocity v of an electron in the band, and the acceleration dk/dt in terms of a force $-dV/dx$, as

$$v = \frac{1}{\hbar}\frac{d\varepsilon_k}{dk}$$

and

$$\hbar\frac{dk}{dt} = -\frac{dV}{dx}.$$

(4.4)

These are exactly the counterparts of Hamilton's Equations with ε_k playing the role of the kinetic energy and $\hbar k$ playing the role of momentum. Thus the bands are the basis of our understanding of most of the electronic properties of solids.

Before going to three dimensions we should seek the solution for a finite chain of atoms, with ends, as shown in Fig. 4.1(b). We note that the solutions we found for the circular chain had complex coefficients but, since the states with $-k$ and k have the same energy we can also write a solution by adding them, giving coefficients $u_j = \cos(kx_j/\sqrt{(2/N)})$. We could also subtract the two, giving $u_j = \sin(kx_j/\sqrt{(2/N)})$. If we now choose such a solution which is zero at one atom distance to the left of a series of atoms, and also at one atom distance to the right, the variational equations Eq. (4.2) are satisfied for every atom in that series, even if the atoms beyond the chain are removed. We have found a solution for this series, this finite chain of atoms. Such solutions require that $k(N+1)d$ be an integral number times π so that the coefficients can be zero just beyond and before the chain. Eq. (4.3) again gives the energy and these could be the solid dots in Fig. 4.2, and dots midway between each set of neighbors, for $0 < kd \le \pi/d$, again one state for every atom. The band is not modified by using these *vanishing boundary conditions* nor can any extra "surface states" arise since there is already one state per atom. This feature will be important when we again consider surfaces of solids.

4.2 Three-Dimensional Lattices

The extension to a simple-cubic lattice, with spacing d, in three dimensions is quite immediate. With a total of N atoms, the coefficients generalize to $\exp(i(k_x x_j + k_y y_j + k_z z_j))/\sqrt{N}$ so that for atomic s states the band states and energy become

$$\psi_k(\mathbf{r}) = (1/\sqrt{N})\sum_j \exp(i(k_x x_j + k_y y_j + k_z z_j))\psi_s\left(\mathbf{r} - \mathbf{r}_j\right)$$

and   (4.5)

$$\varepsilon_{\mathbf{k}} = \varepsilon_s + 2V_{ss\sigma}(\cos(k_x d) + \cos(k_y d) + \cos(k_z d)) + \lambda V_{ss\sigma}^2/|\varepsilon_s|$$

in terms of the three-dimensional vector $\mathbf{k}$ in the cubic Brillouin Zone, with edge $2\pi/d$. We shall not explicitly keep the nonorthogonality term $\lambda V_{sp\sigma}^2/|\varepsilon_s|$, which simply shifts the bands by a constant, but include its effect as needed. There are again N states and for lithium half of them, around the center of the Zone, are occupied for lithium. The shape of the region occupied is

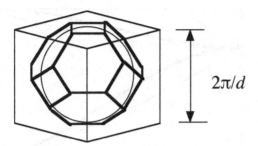

Fig. 4.3. The polyhedron, heavy lines, contains the region of occupied states for lithium within the cubic Brillouin Zone (light lines) of a simple-cubic lattice, for the approximate bands of Eq. (4.5). The sphere of volume equal to the polyhedron, the light-line circle, is the free-electron Fermi surface.

shown in Fig. 4.3 and may surprise solid-state physicists; it is exactly the shape of the familiar Brillouin Zone for a face-centered-cubic lattice. This *Fermi surface* separates occupied and empty states, and with any more complete representation of the states will be rounded, more like the sphere also shown in Fig. 4.3, lying just barely inside the cubic Zone. Eq. (4.5) describes the dynamics for electrons moving in three dimensions in the simple cubic crystal. We can also obtain states for a finite crystal, just as for the finite chain, by applying the vanishing boundary condition a distance d from each surface. The construction of states for the different metal crystal lattices described in Appendix 4B is straightforward, though a little intricate, as might be suggested by the Brillouin Zone for the face-centered-cubic lattice, which happened to be the same as the lithium Fermi surface shown as the polyhedron in Fig. 4.3. This construction of states for such lattices is given in any solid-state text, and is not necessary here.

For metals other than lithium (or other alkali metals) we must include p states as well as s states in the expansion of the wavefunction, Eq. (4.1), just as we included them for the diatomic molecules of those elements. This requires the couplings between p states, $V_{pp\sigma}$ and $V_{pp\pi}$, and the coupling $V_{sp\sigma}$ between s and p states. It also requires the difference in energy $\varepsilon_p - \varepsilon_s$ between them. We shall shortly obtain the corresponding bands just as we obtained the s bands for the linear chain of lithium atoms above. They are the second and third bands shown as heavy lines in Fig. 4.4. The first solid line up, based upon s states looks different from that shown in Fig. 4.2 near the Brillouin Zone because, as we shall see, we included the coupling $V_{sp\sigma}$ between s and pz states.

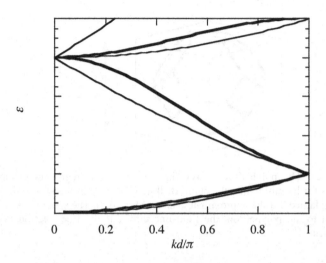

Fig. 4.4. Free-electron bands, shown as light lines, transferred back to the Brillouin Zone, and tight-binding s and p bands, based upon couplings (Eqs. (3.8) through (3.11)), chosen to fit them as closely as possible. Wavenumbers lie in a cube direction from the Zone center.

We would like to plot the energies of free electrons, the states which would occur if the potential arising from the lithium ions were eliminated completely, on the same graph. This requires rewriting the free-electron state energies, $\varepsilon_k = \varepsilon_0 + \hbar^2 k^2/2m$ in the Brillouin Zone for this lattice. We noted in the construction of states for the chain that any states which differed in wave number by $2\pi/d$, or a multiple of it (called *lattice wavenumbers*; in fact the lattice can diffract electrons between any two states differing in wavenumber by a lattice wavenumber if those states have the same energy), were really the same state. Correspondingly, a free-electron state exp($\mathbf{k} \cdot \mathbf{r}$) with $\mathbf{k}$ outside the Brillouin Zone could be regarded as a state with wavenumber translated by some lattice wavenumber back into the Brillouin Zone. Such a contraction of the free-electron states to the Brillouin Zone for a simple cubic lattice is shown in Fig. 4.4. If we wish to compare the resulting free-electron bands with our bands based upon atomic orbitals, the p bands will need to be associated with the higher free-electron bands.

Indeed there is an amazing similarity between the bands such as those from Eq. (4.4) and the free-electron bands as seen in Fig. 4.5. It is of course no accident since we adjusted the couplings and the $\varepsilon_p - \varepsilon_s$ and the ε_0 to make them as close as possible, but it is still impressive how closely they can be fit. The important point is that these two extreme views, electrons localized in atomic states coupled to their nearest neighbors, and electrons completely free, with negligible effect from the atoms present, can be applicable to the same system. We also saw this in Fig. 4.3 where the Fermi surfaces for the two descriptions are qualitatively similar.

For what we attempt here an essential consequence is that for these simple metals we have a very good idea what the couplings must be for the description in terms of tight-binding states based on coupled atomic orbitals. The width of the s band in Fig. 4.4, from Eq. (4.5), is $\varepsilon_{\pi/d} - \varepsilon_0 = -4V_{ss\sigma}$, and for the free-electron band it is $\hbar^2(\pi/d)^2/2m$, from which we obtain $V_{ss\sigma} = -(\pi^2/8)\hbar^2/(md^2)$, which we gave as Eq. (3.8). For the $|pz>$ states we similarly have bands from the counterpart of Eq. (4.5) for $p\sigma$ states as $\varepsilon_k = \varepsilon_p + 2V_{pp\sigma}\cos(kd)$, and from Fig. 4.4 we see that the corresponding free-electron band has three times the width and negative slope so $V_{pp\sigma} = (3\pi^2/8)\hbar^2/(md^2)$, which we gave as Eq. (3.9). Similarly, the π bands correspond to free electron states of energy $\hbar^2[(2\pi/d)^2 + k^2]/2m$ which were transferred in from lateral directions and we see that $V_{pp\pi} = V_{ss\sigma}$ as in Eq. (3.10). Finally, we may see that the coupling between the s-band states and the $p\sigma$-states is $\pm 2iV_{sp\sigma}\sin(kd)$ as described in Appendix 4A. This does not affect the energies at k equal to zero and π/d where we did the matching, but it shifts the slope of the bands at $k = \pi/d$, and if we choose $V_{sp\sigma} = (\pi/2)\hbar^2/(md^2)$, Eq. (3.11), we obtain the free-electron slope at that point as we see in Fig. 4.4, completing our derivation of the sp couplings. The largest remaining discrepancy between the free-electron and tight-binding bands is for the p bands near $k = 0$, and that could be remedied by adding d states and $V_{pd\sigma}$ as we added $V_{sp\sigma}$.

For lithium it might not be so unreasonable to assume that the coupling for the metal might be applicable to the diatomic molecule. We shall see that the energy bands of the semiconductors are also free-electron-like, so we might extend these couplings to apply to all other systems, though slightly different coefficients are obtained with different crystal structures. Application to ionic solids or oxides is much more speculative. We did in fact find sizable deviations for diatomic molecules requiring the scalings in Table 3.1. The coefficients are inevitably approximate in any case. We obtained the $-\pi^2/8$ for $V_{ss\sigma}$, but for the diamond structure of most semiconductors we would obtain $-9\pi^2/64$ and for the face-centered-cubic structure of many metals we would obtain $-\pi^2/12$.

Fitting the bands also requires a choice of $\varepsilon_{p^-}\varepsilon_s$ equal to $\pi^2\hbar^2/md^2$. This is quite close to the values obtained from Table 1.1 for the tetravalent metals, tin and lead, but somewhat above for the trivalent metals (a factor 1.8 for Al, and 1.4 for Tl) and higher for the divalent metals, though we only have extrapolated values for the p states there. The lower $\varepsilon_p - \varepsilon_s$ values obtained from Table 1.1 are in fact responsible for some of the real deviations from free-electron behavior in some simple metals, as we see when we treat lithium metal in detail in Appendix 4A. We may note also that if we had kept the $\lambda V_{ss\sigma}^2 / |\varepsilon_s|$ in Eq. (4.5), and inserted the corresponding term for the p states it would have shifted the p state more than the s state. providing an increase in $\varepsilon_p - \varepsilon_s$, but for our purposes it is best to continue with the atomic levels from Table 1.1.

We may ask how the atomic picture can be so close to free electrons. We noted the clue in Chapter 1 where we replaced the true deep potential arising from an atom by the empty-core pseudopotential in Fig. 1.3 giving the same wavefunction outside the core radius r_c and the same atomic energy. Doing that in a simple-cubic lithium lattice, with the same density of atoms as the true body-centered-cubic lithium, gives the potential along a line of atoms shown in Fig. 4.5. The important point is that it is weak, in the sense that the eigenstates (pseudowavefunctions) are smooth near the nucleus as for $\varphi(r)$ in Fig. 1.3, without the structure associated with the core. Another important feature, which will come up in Section 4.5, is that the pseudopotential is *higher* near that atoms than between. It favors electrons being between the atoms rather than on them.

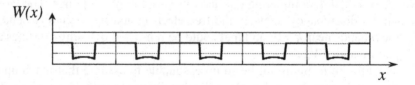

Fig. 4.5. The pseudopotential as a function of position x along a line of atoms in simple-cubic lithium. The faint vertical lines are at the nuclear positions.

4.3 Cohesion in Simple Metals

This view of the simple metals based upon both atomic and free-electron approaches gives us a simple way to estimate the cohesive energy of the metal, which we do again for simple-cubic lithium. We begin with one electron on each atom at energy ε_s, and bring the atoms together with spacing d which forms bands, with the band minimum below ε_s by $6V_{ss\sigma} =$

$-(3\pi^2/4)\hbar^2/md^2$, gaining us that much energy per atom. However, these electrons also have kinetic energy, most accurately obtained as free-electron bands, filled up to a Fermi sphere of radius k_F (in wavenumber space) as in Fig. 4.3. Since half the states in the Brillouin Zone are doubly occupied, the Fermi surface volume is half the Brillouin-Zone volume, $4\pi k_F^3/3 = \frac{1}{2}(2\pi/d)^3$, or $k_F d = (3\pi^2)^{1/3}$. Averaging the kinetic energy over the sphere gives 3/5 the value at the Fermi energy, or $(3/10)(3\pi^2)^{2/3}\hbar^2/md^2 = 2.87\hbar^2/md^2$ per atom. The sum of these two energies drops with decreasing d but if we reintroduce a repulsion, equal to some combination of terms such as $\lambda V_{ss\sigma}^2/|\varepsilon_s|$ between each pair of neighbors, and adjust λ so that the minimum energy occurs at the equilibrium spacing, it will cancel half the attractive energy. We are left with a cohesive energy per atom (again negative for lower energy in the solid) of

$$E_{coh} = \frac{1}{2}\left(-\frac{3\pi^2}{4} + \frac{3}{10}(3\pi^2)^{2/3}\right)\frac{\hbar^2}{md^2} = -2.27\frac{\hbar^2}{md^2}. \tag{4.6}$$

It seems remarkable that such a complicated property could be estimated with such a simple formula. This could also be written in terms of the Fermi wavenumber k_F as $-2.27\hbar^2/md^2 = -0.237\hbar^2 k_F^2/m$. For a face-centered-cubic structure (fcc structure, see Appendix 4B) one similarly obtains $-0.226\hbar^2 k_F^2/md^2$ for a monovalent metal. The coefficients are so close in

Table 4.1. Cohesion in the simple metals, in eV per atom, assuming a face-centered-cubic structure, for a free-electron Fermi wavenumber k_F (Å^{-1}) obtained for the true structure, from Harrison (1999), p. 442.

Mono-valent	k_F	Eq. (4.6)	Exper.	Divalent	k_F	Modified Eq. (4.6)	Exper.
Li	1.11	−1.47	−1.63	Be	1.94	−2.49	−3.32
Na	0.92	−1.01	−1.11	Mg	1.37	−1.24	−1.51
K	0.75	−0.67	−0.93	Ca	1.11	−0.81	−1.84
Rb	0.70	−0.59	−0.85	Sr	1.08	−0.77	−1.72
Cs	0.65	−0.51	−0.80	Ba	0.98	−0.63	−1.90
Cu	1.36	−2.21	−3.49	Zn	1.58	−1.65	−1.35
Ag	1.20	−1.72	−2.95	Cd	1.40	−1.30	−1.16
Au	1.21	−1.75	−3.81	Hg	1.36	−1.22	−0.67

terms of k_F that it is reasonable to use the fcc formula for all monovalent metals, giving the results in Table 4.1. For a divalent metal, such as Mg or Ca, $k_F d$ is increased by a factor $2^{1/3}$ and we add the effect for two electrons per atom, leading to the values also in Table 4.1, and results were similar for trivalent and tetravalent metals (Harrison, 1999). The general agreement seems remarkable, particularly for the alkali metals, and it is interesting that we could stretch it so far. We attribute the larger discrepancies below the first two rows to additional contributions from d states, not included here.

4.4 Other Properties

It is also possible to understand (Harrison, 2000, 26ff) the surface energy in terms of this free-electron picture, as we found the change in energies due to using a finite chain rather than periodic boundary conditions in Section 4.1. We may think of a metallic slab, of thickness L in the z direction, but with periodic boundary conditions in the other two directions. Then $k_z L$ must be an integral multiple n of π to satisfy the vanishing boundary conditions, and the kinetic energies of the states become

$$\varepsilon_k = \frac{\hbar^2 \pi^2 n^2}{2mL^2} + \frac{\hbar^2 (k_x^2 + k_y^2)}{2m}. \tag{4.7}$$

We have summed these carefully (Harrison, 1999, 716ff; 2000, 24ff) to obtain the energy per electron as a power series in $1/L$. The zero-order term is just the $^3/_5 \hbar^2 k_F^2 / 2m$ which entered the cohesion. The term proportional to $1/L$ gives the surface energy,

$$E_{surf.} = \frac{k_F^2 E_F}{8\pi} \left(1 - \frac{8 k_F s}{15\pi} \right), \tag{4.8}$$

where s is the distance between atomic planes in the bulk; note that the result has units of energy per unit area. The first term comes from the sum of free-electron energies at fixed L and the second comes from the fact we noted for the linear chain, that the vanishing boundary conditions should be taken a full lattice distance from the last plane of atoms, giving a dimension $L + d$ (here $L + s$), allowing some relaxation of the electrons. For the simple-cubic monovalent metal $k_F d = (3\pi^2)^{1/3} = 3.09$ and the second term is 0.53 times the first for (100) surfaces. This term again required use of both the tight-binding and free-electron approaches. This also gives a dependence upon which crystallographic plane forms the surface. Terms proportional to $1/L^2$ give the interaction between surfaces (Harrison, 1999, 716ff). They are oscillatory as a function of L and responsible for one form of *Giant Magnetoresistance* (ibid.).

A notable feature of Eq. (4.8) is that the surface energy is of the same order as the cohesive energy per bond, defined for simple structures as the cohesive energy per atom divided by half the number of nearest neighbors. For a monovalent simple-cubic metal, again $k_F d = (3\pi^2)^{1/3}$, and the leading term in Eq. (4.8) becomes $[(3\pi^2)^{4/3}/16\pi]\hbar/md^4$, corresponding to 1.82 $\hbar^2/md^2$ per broken bond (one bond per d^2 of surface). Dividing the corresponding cohesive energy of Eq. (4.6) by half the six neighbors gives a comparable 0.72 $\hbar^2/md^2$. Interestingly enough, we will find that this plausible relation between surface energies and number of bonds broken is completely *inappropriate* for covalent and ionic crystals. It is approximately true only for simple metals, seemingly the least likely case.

For electronic properties of the simple metals, the nearly-free-electron view alone is much simpler and more effective than the tight-binding view. In Harrison (1999) we constructed the matrix elements between plane-wave electron states in terms of the empty-core pseudopotential described in Section 1.4, and included screening of these by the redistribution of charge (in the Fermi-Thomas Approximation). We could then treat the true, deformed Fermi surfaces, scattering of electrons by defects and impurities, and the interaction between electrons and phonons, all in terms of the empty-core radii listed in Table 1.1. We do not repeat these here.

We also used these pseudopotentials to find the change in total energy of the lattice due to lattice deformations and could calculate the vibration spectrum itself and a number of properties associated with defects, alloying and liquid metals. One very interesting aspect was that it was possible to write the total energy in terms of two-body interactions between atoms a distance r apart,

$$V_0(r) = \frac{Z^2 e^2 \cosh^2(\kappa r_c) e^{-\kappa r_c}}{r}, \tag{4.9}$$

where $\kappa^2 = 4e^2 mk_F/(\pi\hbar^2)$ is the square of the Fermi-Thomas screening constant. This is in addition to other terms in the energy which depend upon total volume, but not on the detailed arrangement of atoms. This result is counterintuitive; we are used to thinking of interactions which have a minimum near the equilibrium spacing so that the system can be in equilibrium with only those interactions. However, we regard this new result as essentially correct. If the system *were* in equilibrium under central-force interactions alone, then the Cauchy relation between the elastic constants, $c_{44} = c_{12}$ for cubic crystals, would obtain and they are experimentally far from satisfied. We may use Eq. (4.9) to calculate properties, such as vibration spectra, at constant volume. Then elastic constants (including even the bulk modulus) can be deduced from them. Indeed they are found not to satisfy the Cauchy relations and roughly explain the deviations from them.

4.5 Surface States and H_2O on a Metal Surface

For the linear chain with s states, we found in Section 4.1 that truncating the crystal left the energy bands the same, just shifted the wavenumbers at which the eigenstates occurred within those bands. If we changed the energy level for the end atom sufficiently (a large shift of $\pm V_{ss\sigma}$ is needed) we could form a surface state, a state with tight-binding coefficients u_j which decay as $\exp(-\mu dj)$ away from the end. The state has been pulled out of the band, since the total number of states remains equal to the number of atoms, but their energy lies beyond the band edge. [It is much easier to create such a state by shifting a level on an atom not near the ends. Slightly lowering the energy level of an atom far from the ends always pulls a level out of the bottom of such a one-dimensional band, an *impurity* state which decays in both directions away from the atom with the shifted level. In a three-dimensional crystal the shift of the impurity level must exceed some critical value.] We can make the same statement for isolated bands based upon p states. [For Fig. 4.4 the s and p bands touched at the zone face, but they would separate if we increased ε_p above the value chosen for that plot. Then adding a $V_{sp\sigma}$ would again push the bands further apart, as it did for Fig. 4.4, making them more free-electron-like.]

Something novel and different happens if these isolated bands cross each other, as by lowering the ε_p from that used for Fig. 4.4. This is in fact the usual case as we find when we construct the tight-binding bands for lithium in Appendix 4A. There, such a crossing of bands is illustrated in Fig. 4A.1. We can also see why this crossing occurs in terms of plane-wave states and the pseudopotential of Fig. 4.5. At the Zone boundary, $kd = \pi$, the s-like state is a cosine wave, with maximum at the nuclear positions where the pseudopotential is high. The p state is a sine-like wave, zero at the nuclear positions, and maximum between the nuclei, where the pseudopotential is low, as we noted in discussing that curve. Thus the pseudopotential will shift the p-like state below the s-like state. It was pointed out by Shockley (1939) that in this case the bands are then separated for all wavenumbers and two surface states appear in the middle of the gap.

In hindsight it is easy to see why this happens. We can construct hybrid states on each atom $(\psi_s \pm \psi_{p\sigma})/\sqrt{2}$ just as we did for the σ states in the CN molecule in Chapter 3. Then between each pair of atoms we construct bonding states with the inward-leaning hybrids. The residual coupling between neighboring bonds, and between neighboring antibonds, broadens them into bands not so different from the bands formed by isolated atomic states. [We shall see in the next chapter that this formation of bonding and antibonding bands is the essential feature of most semiconductor bands.] However, there is a leftover *dangling hybrid* at each end, pointed outward, which becomes a localized state at each end. These Shockley surface states are distinguished from those obtained with a large shift of level of the end atoms, which are called Tamm states (Tamm, 1932).

We may construct a three-dimensional crystal by combining such chains, with the end atoms forming the surfaces of the crystal. These dangling-hybrid states are coupled to each other at each surface forming *surface bands* (described for example in Harrison (2003)). The presence of surface bands will affect the surface energy significantly, as we shall see in Chapter 5 for semiconductors.

We may now return to the problem of a water molecule above a simple-metal surface. We considered the effects of the image potential in Chapter 2, which are valid contributions to the energy. But we may also consider bonding interactions such as we discussed in Chapter 3. In order to do this we need to consider the states of the water molecule and the metal more completely, which we do in Appendixes 4A and 4B. We did this for water with the angle between protons as $90°$ as we found in Chapter 3, and we treated lithium as a simple cubic metal. This may be a very representative case of how we can use the understanding of the electronic structure and molecules to present an approximate description of a complicated situation. We obtain in Appendix 4B the energy levels shown in Fig. 4.6.

On the left are the energy levels for the water molecule. There are two antibonding levels at higher energy, which are empty, and two bonding

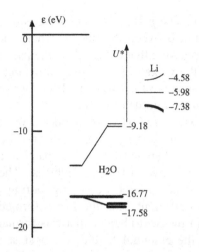

Fig. 4.6. Energy levels for water, and the bands near the Zone boundary for simple-cubic lithium. That at −5.98 eV is the dangling hybrid surface band. Occupied states are drawn heavy, empty states light.

levels at lower energy which are occupied. The third oxygen p state, perpendicular to the first two, is uncoupled and occupied. For the lithium we show the bands near the Zone face. We note that the occupied levels in the lower lithium band lie above the empty antibonds on the water, which would suggest incorrectly that electrons should be transferred. However, we noted that for adding an electron to the neutral molecule, this should be the electron affinity, above the levels shown to which an electron *within* the molecule would be transferred from a bond level. This adds an energy U (which we take to be $e^2/d = 15$ eV in terms of the hydrogen-oxygen distance d), corrected by the energy $-e^2/4z$ by which it would be reduced by the image potential from the substrate, for an effective $U^* = U - e^2/(4z)$. This is enough to raise the affinity level for the molecule well above the lithium occupied levels down to $z = 0.3$ Å. The dangling hybrid level from the lithium is essentially an outward-leaning hybrid and will be much more strongly coupled than the band states, based upon inward-leaning hybrids, as we saw earlier in this section. Thus the dominant contributing coupling will be between this empty dangling hybrid and the occupied oxygen bonds, far below in energy (the coupling of the empty hybrid to the empty antibonding state does not affect the total energy). With this large energy difference we do not expect the energy shift to be important and so for this monovalent metal the electrostatic contributions we discussed in Section 2.5 may well be dominant.

For a divalent metal, the Fermi energy will lie above the energy of the dangling hybrid and it will be occupied. [In a divalent metal there are always electrons in the second band; that band is lowest at this Zone face, and the hybrid is even lower.] Then the dominant coupling is between the occupied dangling hybrid and the empty antibonding state, at an energy which comes down towards the dangling-hybrid energy as the molecule comes close and U^* decreases. This one could be important, and the most interesting question is whether it favors the protons in the water being toward the metal, or hanging out away from the surface. We did not attempt a full calculation, but since it is the same antibonding state of the water, which enters for either geometry, we simply estimated the strength of the coupling between the dangling hybrid and the antibonding state, as a function of the distance z to the nearest atom. We found a coupling of $V_2 =$ 2.09 $\hbar^2/mz^2$ if the oxygen ion was closest and 2.89 $\hbar^2/mz^2$ if the hydrogen ion was closest, suggesting that the orientation of the proton *toward* the metal surface is favored, as we found for the electrostatic interactions in Section 2.5. Note that the relative value is unaffected if we introduce a scale factor as from Table 3.1 and that adding a repulsion $\lambda V_2^2/\sqrt{(\varepsilon_i\varepsilon_j)}$ still leaves the stronger coupling dominant. We conclude that this coupling leaves our conclusion the same, contrary to the long-standing paradigm and consistent with what Feibelman (2010) indicates is the current status for water on metals such as ruthenium, as we discussed in Section 2.5. Ruthenium is a transition metal so our estimates are not directly applicable. d states should be included in the substrate, as we do in Chapter 7, and we would anticipate that an occupied d orbital would play the role of the dangling-hybrid lithium state, leading to the same conclusion, but we have not carried out that study.

These arguments are far from certainties, but they are much better than the simple chemical arguments based upon such quantities as the relative strength of a lithium-oxygen bond and a lithium-hydrogen bond, which are only indirectly relevant. Our arguments are also much easier to make and to understand than full density-functional calculation.

sulphur

CHAPTER 5

Covalent Solids

We noted in Section 1.2 that most of the elements in Column IV of the Periodic Table form covalent solids (though lead, at the bottom, is a simple metal and tin can be either). As we shall see, these covalent solids have small gaps separating the occupied states from the empty states, making them semiconductors; carriers will be present from thermal excitation across the gap. As a crystal each atom has four nearest neighbors, arranged as the corners of a regular tetrahedron, as shown in Fig. 5.1. The structure is the diamond structure, but we prefer to think in terms of silicon. They are generally viewed as placing one electron from each atom in each of its four surrounding two-electron bonds, and we also take this view, reducing the problem of a complete solid to a sum of nearly independent bonds. It is accomplished by making hybrid s and p states just as we did for the linear chain of lithium atoms to demonstrate surface states, in Section 4.5. We shall then construct bands and see how they contrast to simple-metal bands.

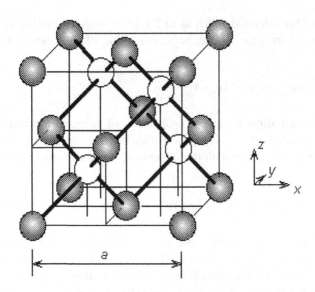

Fig. 5.1. The zincblende structure, which becomes the diamond structure if both the shaded (zinc) atoms and the open (sulfur) atoms are replaced by carbon, or by silicon. a is the cube edge. The nearest-neighbor distance is $d = (\sqrt{3}/4)a$.

5.1 Homopolar Semiconductors

For the linear chain we transformed from s states and p states aligned with the chain; with one of each these needed to be $(|s> \pm |p>)/\sqrt{2}$, normalized and orthogonal to each other. For the tetrahedral solid, for an atom with nearest neighbors in the directions [1,1,1], [1,−1,−1], [−1,1,−1], and [−1,−1,1] we may similarly construct the natural generalization as normalized, orthogonal hybrids,

$$|h_1> = (|s> + |x> + |y> + |z>)/2,$$

$$|h_2> = (|s> + |x> - |y> - |z>)/2,$$

$$|h_3> = (|s> - |x> + |y> - |z>)/2, \tag{5.1}$$

and

$$|h_4> = (|s> - |x> - |y> + |z>)/2,$$

where we have written the p state oriented along the x axis as $|x>$, etc. The nearest-neighbor atoms have their neighbors in four exactly opposite

directions and have hybrid states as in Eq. (5.1) with the coefficients of each
p state reversed in sign. The expectation value of the energy for each of
these is

$$\varepsilon_h = <h_j|H|h_j> = (\varepsilon_s + 3\varepsilon_p)/4. \qquad (5.2)$$

We see that it has three times the probability of being in a p state than an s
state so it is called an sp^3 hybrid. It clearly leans in the direction of the
neighbor and the coupling with the opposing hybrid on that neighbor is quite
strong,

$$<h_j|H|h_i> \equiv -V_2 = (-3V_{pp\sigma} - 2\sqrt{3}V_{sp\sigma} + V_{ss\sigma})/4 = -3.86 \; \hbar^2/md^2, \qquad (5.3)$$

in comparison to the coupling with one of the other hybrids on that atom,
which is $<h_j|H|h_i> = (V_{pp\sigma} - (\sqrt{3} - 1/\sqrt{3})V_{sp\sigma} + V_{ss\sigma})/4 = 0.16\hbar^2/md^2$ (the cosine
of the angle between two hybrids on the same atom is $-1/3$). Thus it makes
sense to take this $-V_2 = -5.33$ eV for silicon with $d = 2.35$ eV as the
dominant coupling in the problem, forming bonds and antibonds between
each pair of neighbors with energies

$$\varepsilon_b = \varepsilon_h - V_2,$$

and $\qquad\qquad\qquad\qquad\qquad\qquad\qquad\qquad\qquad\qquad\qquad\qquad (5.4)$

$$\varepsilon_a = \varepsilon_h + V_2.$$

It is interesting to pause here to see what this suggests for the cohesion
in covalent solids such as silicon. Each atom initially had two electrons at ε_s
and two electrons at ε_p and after forming the crystal all four electrons are in
the bond state with ε_b. At the minimum energy, the nonorthogonality shift
$\lambda V_2^2/|\varepsilon_h|$ will equal $V_2/2$, as we have found before. Thus the predicted
change in energy in forming the solid, per atom, is

$$E_{coh} = 4\varepsilon_h - 2V_2 - 2\varepsilon_s - 2\varepsilon_p = \varepsilon_p - \varepsilon_s - 2V_2. \qquad (5.5)$$

The term $\varepsilon_p - \varepsilon_s$ is called the *promotion energy*, which we encountered when
we made hybrids before. Here it is the energy required to prepare the atoms
in the sp^3 configuration, 7.20 eV for silicon. The remainder is the bonding
term, analogous to Eq. (4.6) for simple metals, equal to -10.65 eV for
silicon. The difference is -3.45 eV, to be compared with the experimental
-2.32 eV, not so bad as the difference between large values. A treatment of

cohesion in the entire range of covalent solids, using couplings tuned to semiconductor bands, is given in Harrison (1999).

A most important feature is that the cohesion *is* the small difference between sizable contributions. If we were to create a surface to the crystal, otherwise holding all atoms in their initial positions, we would again form sp^3 hybrids, at the cost of the promotion energy, but at the surface we would have dangling hybrids, as we did for the metal surface. Then we fail to gain the −5.33 eV for that broken bond. Thus the surface energy is 5.33 eV per broken bond, very much larger than the experimental cohesive energy of 1.16 eV per bond. As we mentioned before, the use of the energy per bond as an estimate of surface energy is qualitatively correct for simple metals, which have no promotion-energy contribution, but it is not even qualitatively correct for covalent solids where the description in terms of bonds seems otherwise quite sensible, because neither requires promotion energy.

This feature of a large surface energy in comparison to the cohesion has another consequence. It causes many surfaces of covalent solids to *reconstruct*, rearrange the atoms at the surface. The system seeks a way to recover some of this lost promotion energy. One way it can do that is for atoms with doubly occupied dangling hybrids to move out from the surface, causing the angles between the hybrids to change, and the hybrids to change from sp^3. Outward motion makes the dangling hybrid more *s*-like and with two electrons some of the promotion energy is recovered; similarly atoms with empty hybrids tend to move inward. Another way is to displace the surface atoms sufficiently that they form additional bonds. Such reconstructions tend not to occur in the simple metals, nor in the ionic solids we discuss in Chapter 6.

5.2 Compound Semiconductors

This simple picture of two-electron bonds extends by theoretical alchemy directly to the covalent compounds formed between nonmetals and simple metals to the left of Column IV. As we described in Chapter 1, we may start with silicon and transfer a proton between each pair of nearest neighbors to form AlP, and then MgS, forming the zincblende structure of Fig. 5.1. We construct the same hybrids on each atom, but now the two hybrid energies obtained from Eq. (5.2) have different energy, and a polar energy is defined in terms of the difference,

$$V_3 = (\varepsilon_h^{+} - \varepsilon_h^{-})/2 , \qquad (5.6)$$

with ε_h^+ the hybrid energy for the metallic atom, that which gave up the proton, and ε_h^- is for the nonmetal which received the proton. The bond and antibond energies of course become

$$\varepsilon = (\varepsilon_h^+ + e_h^-)/2 \pm \sqrt{(V_2^2 + V_3^2)} + \lambda V_2^2 / \sqrt{(\varepsilon_h^+ \varepsilon_h^-)}. \tag{5.7}$$

The theory of the cohesion proceeds exactly as for silicon, with promotion energy different for the two atoms, and an addition term transferring an electron from ε_h^- to ε_h^+ so one electron is in each hybrid. It is interesting that with this additional promotion, the cohesive energy decreases with increasing polarity. Somehow it has become common to think of this backwards, as there being an extra "ionic contribution" to a chemical bond. Perhaps the reason is that the *formation energy*, the energy to separate the compound into its component elements, *will* increase with polarity. There is a formation energy to separate AlP into Al and P but, by definition, the energy per atom to separate a large piece of silicon in its standard state into two pieces of silicon in the standard state is zero, except for surface energy which per atom is negligible in comparison to the bulk energy. The cohesive energy is a more relevant measure of bonding energy.

We can again define a polarity of the bonds, $\alpha_p = V_3 / \sqrt{(V_2^2 + V_3^2)}$ as in Eq. (3.20) and write the coefficients for the states themselves in terms of it as in Eq. (3.21). These bond and antibond orbitals replace the molecular orbitals. This simplest picture of the electronic structure of covalent solids provides a way to estimate a wide range of properties, as in the other systems we have considered. The bond polarizability calculated this way indicates correctly a very large dielectric constant but does not predict it very accurately. This polarizability leads to a dielectric susceptibility of (Harrison, 1999, p. 147)

$$\chi = \frac{\sqrt{3} e^2 \alpha_c^3}{8 V_2 d} \tag{5.8}$$

in terms of a covalency given by $\alpha_c = V_2 / \sqrt{(V_2^2 + V_3^2)}$, in analogy with the polarity defined in Eq. (3.20). This covalency is one for the homopolar covalent elements and drops with the polarity of the compounds. The dependence upon polarity in the semiconductors is well given by this form, but the values are too small, increasingly so for the heavier elements. This difficulty arises partly from the coupling between bonds and neighboring bonds and antibonds, which also converts the bonding and antibonding levels into valence bands and conduction bands, as described in Section 5.3.

Similarly we may estimate the rigidity of the lattice, calculating the change in energy under a lattice distortion which tends to twist the neighbors from tetrahedral toward being in a single plane. This leads to a predicted elastic shear constant (ibid. p. 113)

$$c' = \frac{c_{11} - c_{12}}{2} = 2.38 \frac{\hbar^2 \alpha_c}{md^5}. \tag{5.9}$$

This gives quite a good account of the measured values for the homopolar semiconductors, but the experimental dependence upon polarity is more like the α_c^3 of Eq. (5.8), which we found also to be due to neglected terms. In Harrison (1999) we considered a very wide range of such properties, but in term of the simplest representation of bond orbitals, and including the intra-atomic terms which we referred to as *metallization* of the bonds in Section 3.4 and in Appendix 3D4. Here we look only at the production of energy bands, and the insight into the bonding which it gives.

5.3 Energy Bands

The intra-atomic coupling is the coupling between two hybrids on the same atom, as for molecules, which is readily evaluated for the tetrahedral solid as

$$<h_j|H|h_j> \equiv -V_1 = (\varepsilon_s - \varepsilon_p)/4. \tag{5.10}$$

(In terms of it the cohesive energy per bond was $-V_2 + 2V_1$.) For the homopolar compounds the coupling between neighboring *bonds* is reduced, by the coefficient of that hybrid on each, to $-V_1/2$ (or to $-(1 \pm \alpha_p)V_1^{\pm}/2$ for the polar semiconductors with the V_1 evaluated for the atom in question).

It is not difficult to construct the bands arising from coupled bond orbitals for wavenumbers along a cube axis for the structure shown in Fig. 5.1 for the homopolar case (ibid. p. 187). We write Bloch sums (Eq. (4.1) with $u_j = e^{ikx}{}_j/\sqrt{N}$) for each of the four orientations of bond orbital, and note that each Bloch sum is coupled to two other Bloch sums of the same wavenumber by $-V_1 \cos(ka/4)$ (with a the cube edge in Fig. 5.1), and to the other Bloch sum by $-V_1$. This gives four equations in the u_j with enough symmetry to solve them together easily, leading to two bands independent of wavenumber at $\varepsilon_k = \varepsilon_h - V_2 + V_1 + V_2/2$. And two others given by

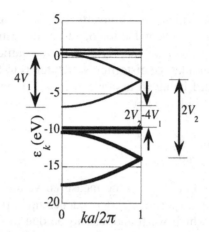

Fig. 5.2. The simplest form for the energy bands along a [100] direction for a homopolar semiconductor based upon the parameters evaluated here for silicon. Occupied bands are drawn heavier. Two close parallel bands are degenerate, and independent of k.

$$\varepsilon_k = \varepsilon_h - V_2 - V_1 \pm 2V_1 \cos(ka/4) + V_2/2. \tag{5.11}$$

(the final term is from nonorthogonality). For the antibonding states, we can change the sign of every hybrid on the light-colored atoms in Fig. 5.1, the couplings on every atom are the same as for bonds and we obtain Eq. (5.11) with $-V_2$ replaced by $+V_2$, These bands are shown in Fig. 5.2.

These crude bands are informative. The gap between occupied and empty states is quite small on the scale of the band widths, $2V_2 - 4V_1 = 3.46$ eV with our parameters, comparable to the observed optical threshold of 4.18 eV. The true conduction band, shown in Fig. 5.3, bends down to the right, rather than rising, giving a smaller minimum gap of 1.13 eV, as is characteristic of a semiconductor. The simplest bands are qualitatively the same as the true bands, with differences coming from neglected couplings.

The *full* tight-binding bands, based upon these same atomic orbitals but slightly different couplings, are shown in the first panel of Fig. 5.3. The states at the bottom of the conduction and the valence bands are purely s-like and given at $k = 0$ by $\varepsilon_s \pm 4V_{ss\sigma}$ (if we drop the repulsive term) which with our parameters would be at -21.60 eV and -7.98 eV, in much better accord with the true bands. Our free-electron fit for the couplings would place the $k = 0$

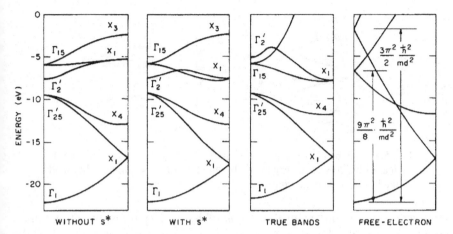

Fig. 5.3. In the first panel are the bands for silicon obtained from a direct tight-binding calculation with parameters similar to ours. In the second panel the effects of a peripheral excited s-state are included. In the third are more accurate bands obtained by Chelikowsky and Cohen (1976), and in the final panel are free-electron bands. The zero of energy in the third and fourth panels was selected so that the valence-band maximum comes at the same energy. The figure is taken from Harrison (1981).

p-like states at $\varepsilon_p \pm 4(V_{pp\sigma} + 2V_{pp\pi})/3$, giving $\varepsilon_p = -5.32$ eV and -9.85 eV. In Harrison (1999) we adjusted the couplings from the free-electron values to accord better with the energy bands of silicon and germanium. Then the full tight-binding bands obtained were those shown in the first panel of Fig. 5.3. Including an additional higher s state, called s^*, allowed us to duplicate the downturn of the conduction band in the second panel. The remarkable accord between the true bands and the free-electron bands is seen by comparison of the third and fourth panels. We may note in passing that the fit of the semiconductor bands to the free-electron bands gives a $V_{ss\sigma} = -(9\pi^2/64) \, \hbar^2/md^2$, very close to the $-(\pi^2/8) \, \hbar^2/md^2$ which we obtained by fitting the simple-cubic bands. If we are interested in properties depending upon the details of the bands, it is certainly necessary to adjust parameters to make it right, but the qualitative bands of Fig. 5.2 may provide a helpful starting point.

In particular, it may be interesting to see how these bands evolved from atomic states if silicon atoms were assembled from wide spacing; this will be representative of how they would evolve also for more complete

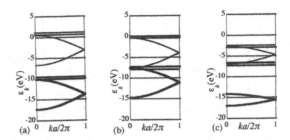

Fig. 5.4. The bands of Fig. 5.2 in part (a). In part (b) the spacing is increased by 20%, and in part (c) by another 20%, becoming s-like and p-like bands, and conducting.

treatments. If we start from the bands of Fig. 5.2, pulling the atoms apart simply reduces the V_2 as $1/d^2$, while leaving the atomic levels and V_1 the same. As we see in Fig. 5.4, the valence and conduction bands move closer together until the gap goes to zero; then the doubly degenerate valence bands attach to the conduction band and a new gap opens up. This leaves two s bands (for the two atoms per primitive cell) below, and six p bands, partly occupied, above. The system has become metallic at larger spacing.

Returning to the bands of Fig. 5.3, we should note also that the distinction between electron affinities arising from the Coulomb U, which we discussed at the end of Chapter 4, also shows up here. The states in the conduction band from the simple estimates we make, and which density-functional theory also gives, are empty states, to which we may add electrons. We might expect that we should add this Coulomb U to those energies. However, this U is reduced by the dielectric constant (ibid, 207ff) to $U/\varepsilon = 7.64$ eV/$12 = 0.64$ eV for silicon. These were included in the bands shown in the third panel in Fig. 5.3 by adjusting the pseudopotentials which were used to fit the observed band gaps (empirical pseudopotentials). Our feeling is that it is preferable not to include this in the energy bands, so that they represent the structure which describes a wide range of properties, and then add this correction if we specifically want the energy for excitations. This same U/ε gives the band-gap enhancement not only for semiconductors, but also for ionic crystals and even rare-gas solids (ibid.).

We have discussed in detail a wide range of properties of semiconductors in terms of this tight-binding approach in Chapters 2 through 8 of Harrison (1999), and will not repeat them here. We should note however that for any discussion of molecules on the surfaces of covalent solids the role of the dangling-hybrid states, which we discussed for metals at the end of the last chapter, will likely be important.

5.4 Other Nonmetallic Elements

Column IV seems designed for covalent solids: four orbitals (one s and three p's) so four sp^3 hybrids for four neighbors, with two-electron bonds. It actually works also for Column V (Harrison, 2002). The s state has become deeper and can be disregarded. With the three p states we can form $p\sigma$ bonds with three neighbors, at $90°$ so the orbitals are orthogonal to each other, leaving the s states doubly occupied. Indeed the structures of black phosphorus, arsenic, antimony and bismuth in Column V are characterized by three nearest-neighbor atoms, with angles slightly larger than $90°$, presumably from the effects of the s states (Harrison, 2002). This also suggests that the cohesive energy per atom would be equal to the lowering of each of the three p electron's energy by $-V_{pp\sigma} = -(3\pi^2/8)\ \hbar^2/md^2$, but reduced by a half by the repulsive interaction as in other systems, $E_{coh} = -3/2V_{pp\sigma}$. Note that in this case there is no promotion energy so the full bonding of the three electrons per atom obtains. This is in moderate accord with experiment, as indicated in Table 5.1, but suggests that our free-electron estimates of the coupling are too large, as they were for the diatomic molecules treated in Section 3.3, and by a similar factor to those given in Table 3.1 (thus closer to a factor ½ as in the heavier atoms, Br_2 and I_2, than the 1/3, or ¼ for the lighter elements).

For Column VI elements there is an additional p electron so we can still make perpendicular σ bonds with two neighbors, but the remaining p state is doubly occupied and cannot form a bond. We expect zigzag chains of atoms and indeed the hexagonal selenium and tellurium structures are aggregates of just such chains (wound as a helix), and sulphur can form an

Table 5.1. Cohesion in Column V and VI Elements

Element	$d(\text{Å})^a$	$-3/2\,V_{pp\sigma}$ (eV)	E_{coh} (eV)^a
P	2.18	−8.90	−3.43
As	3.16	−4.23	−2.96
Sb	2.91	−5.00	−2.75
Bi	3.07	−4.49	−2.18
Column VI		$-V_{pp\sigma}$	
Se	2.32	−5.23	−2.25
Te	2.86	−3.44	−2.23
Po	3.3.4	−2.52	−1.50

^{a}Kittel (1976)

analogous structure. For the cohesion we expect to gain the energy for two electrons per atom in forming the bond, $-2V_{pp\sigma}$, minus a repulsive energy of half that, for a net $-V_{pp\sigma}$ per atom. In Table 5.1 we compare these estimates with experiment, finding that we again overestimate the cohesion, but by a factor of about two. We expect the chains to be held together by much smaller van-der-Waals forces.

In Column VII, with one more p electron per atom, we can only bind with a single atom, and we have seen that the halogen atoms form diatomic molecules in Chapter 3. In Column VIII, the inert gas atoms can form no bond, but are held together with van-der-Waals forces as in Section 1.5.

5.5 Resonant Bonds

Even for Column IV there are alternative structures and alternative electronic structures, graphite being the principal case. The electronic structure of graphite is related to that of the molecule benzene, and we consider the molecule first. It consists of a regular hexagon of carbon atoms, each with a single hydrogen neighbor in the same plane as in Fig. 5.5. We may form covalent bonds between the carbon atoms just as we did for diamond, but with $120°$ angles between neighbors these must be sp^2 hybrids, $|h> = (|s> + \sqrt{2}|p>)/\sqrt{3}$ if they are to be orthogonal to each other and the third hybrid is in the direction of the proton. We can think of the proton as seeking its position of lowest electrostatic energy in the resulting nitrogen-like charge cloud as in Chapter 2, or as forming a bond between the hydrogen and carbon as in Section 3.7. In either case, we are left with one p state oriented perpendicular to the plane, not used in the σ bonds formed, and one extra electron per atom.

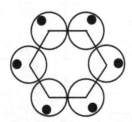

Fig. 5.5. The benzene ring of six carbon atoms in a plane, each with one hydrogen atom in the same plane.

These six states, with π orientation with respect to their neighbors form exactly the kind of chain we introduced for simple metals at the beginning of Chapter 4. They form Bloch states with energy

$$\varepsilon_k = \varepsilon_p + 2V_{pp\pi}\cos(kd) \tag{5.12}$$

with $6kd = 2\pi$ times an integer if the coefficients u_j are to be single-valued. The corresponding band is shown in Fig. 5.6, the counterpart of Fig. 4.2. The three lowest states are doubly occupied, and the upper three empty. With its $d = 1.40$ Å we have $V_{pp\pi} = -4.80$ eV and we gain a bonding energy of $8V_{pp\pi}$, or $4/3\,V_{pp\pi}$ per π electron. If we were to break one bond, applying vanishing boundary conditions as in Section 4.1, we would have doubly occupied states at $kd = \pi/7$, $2\pi/7$, and $3\pi/7$, gaining only $6.99V_{pp\pi}$, lower by $1.01V_{pp\pi}$, so breaking a bond costs less than its original contribution to the bonding. This has traditionally been described as three doubly occupied *resonant* bonds, resonating between different bond sites, a concept introduced by Linus Pauling (e.g., Pauling, 1960). He also introduced an empirical *resonance energy*, added to correct the bond energy. It would seem better to describe it in terms of these global "metallic" states, though with $2V_{pp\pi}$ between occupied and empty states it is more like an insulator. Molecules arise which have multiple hexagons, *aromatic compounds*, sharing one or more sides; naphthalene is a combination of two rings. There are also a vast array of aromatic compounds with one or more of the hydrogen atoms replaced by some other molecular group, or a carbon replaced by other elements; the benzene molecule with silicon replacing the carbon is called silicazine,

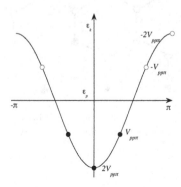

Fig. 5.6. The energy band for π states in benzene. The solid dots are doubly occupied states, the empty dots are empty states.

Graphite is an extension of the aromatic compounds, with atoms in planes having three-fold coordination as shown in Fig 5.7, and no hydrogen. Only recently has it become possible to produce isolated planes, call graphene. In graphite they are stacked, held together predominantly by weak van-der-Waals forces. The electronic structure may be understood in the same way as for benzene. sp^2 hybrids are again formed on each atom, in this case all three forming σ bonds with a neighboring carbon. They broaden into occupied bonding bands and empty antibonding bands with a large gap between them. The π states again form bands as in Fig. 5.6, but now in two dimensions. One might have expected that they would form a simple metal with half-filled bands, but the band structure turns out to be intricate. Because the two shades of atoms in Fig. 5.7 are not translationally equivalent, there are *two* atoms per primitive cell, and two bands, which would be separated by an energy gap if the two atoms were different, as in a boron nitride plane. However, because there is a reflection symmetry which takes one atom into the other, the two bands have *points of contact* between them. This is shown in Fig. 5.8, where the lower band is sketched. Because there is no gap this is a *semimetal*, with the peculiarity that the energy ε_k varies linearly with wavenumber, measured from the point of contact, just as the energy of photons varies linearly with wavenumber. The electronic properties deriving from such bands in graphene are currently an active research area, but will not be pursued here.

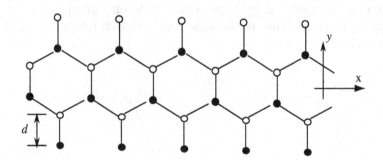

Fig. 5.7. The structure of a graphite plane. All bond angles are 120°. The differently shaded atoms are not translationally equivalent.

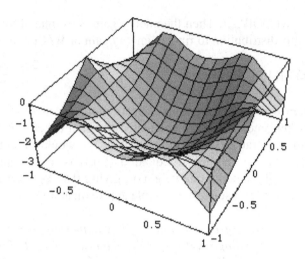

Fig. 5.8. The lower of the two energy bands for the π-states in graphite obtained from an analytic form (Eq. (20-15) in Harrison (1999)). Peaks occur at the corners of the hexagonal Brillouin Zone, inscribed in this box. An identical band (empty) is the reflection of this band through the top plane of the box.

An interesting method for treating the bonding associated with these π bands is the *moments method* (e.g., Harrison, 1999, 93ff) which we have mentioned before, and which we will find useful also for treating d- and f- shell metals. The bands are derivable through the expansion of the wavefunctions in one π-state for each of the N atoms in the crystal, each coupled to three neighboring π states by $V_{pp\pi}$. This is equivalent to diagonalizing a Hamiltonian matrix H_{ij} with every diagonal element $H_{ii} = \varepsilon_p$ with $V_{pp\pi}$ appearing in off-diagonal positions. The second moment of the solutions, measured relative to ε_p can be written (ibid.)

$$M_2 = \frac{1}{N}\sum_{ij} H_{ij}H_{ji} = XV_{pp\pi}^2 , \qquad (5.13)$$

with $X = 3$, the number of nearest neighbors. We might model the density of electronic states per atom per unit energy $n(\varepsilon)$ by a square distribution. $n(\varepsilon)$ $= 1/W$ for $-W/2 < \varepsilon - \varepsilon_p < W/2$, called the Friedel Model (Friedel, 1969). This square distribution has a second moment of $M_2 = W^2/12$. Thus we can obtain a suitable width for the model distribution by equating this to Eq. (5.13), obtaining $W = \sqrt{(12X)}|V_{pp\pi}|$. Then the energy gain is estimated by filling the lower half of this distribution to find an energy gain of $W/4$ per atom, or

$$\delta E = \sqrt{(3X/4)}\, V_{pp\pi} \text{ per electron}. \tag{5.14}$$

It is interesting to note that this approximate moments method could have been used for benzene, where we calculated the gain exactly as $(4/3)\,V_{pp\pi}$, with the factor 1.33 rather than the factor 1.22 we would obtain from Eq. (5.14) with just $X = 2$ neighbors. Equation (5.14) is even approximately correct for the single π bond, one neighbor, giving $0.87\,V_{pp\pi}$ rather than the exact $V_{pp\pi}$ per electron. It suggests a simple approximate rule:

The gain in energy of a resonant bond is approximately equal to the energy of the corresponding single bond times the square root of the number of bond sites between which it resonates.

On one of the few occasions when the author had a chance to discuss with him, he asked Pauling if he knew of the suggestion that the resonant bond had an energy proportional to the square root of the number of resonant sites. He thought a moment, and said it is approximately true. The answer was interesting in two ways: first, that he had such a command of the numbers that he could check it in his head on the spot. Second, for perhaps the same reason, he did not ask nor seem to care how the suggestion arose.

In graphite the resonant bond makes a significant contribution to the cohesion because it is not partially canceled by a promotion energy. With the three sigma bonds it provides a slightly lower energy than the four sigma bonds of the diamond structure. The energy of the graphite structure is higher than that for a tetrahedral structure for silicon, germanium, and tin.

5.6 Covalent Insulators

The dominant insulator in modern electronic chips is silicon dioxide, which can also be regarded as a covalent solid. It also appears as quartz and is the principal component of window glass. It forms a structure with each silicon atom surrounded by a regular tetrahedron of oxygen atoms, a distance $d = 1.61$ Å away, and each oxygen between two silicon atoms, at an angle of approximately $144°$ — in glass as well as in quartz. Pantelides and Harrison (1976) studied the bonding by forming sp^3 hybrids on the silicon, considering a bonding unit of an oxygen atom and two neighboring hybrids, initially ignoring the $-V_1$ which couples the hybrids on the same silicon entering different bonding units. It is in the same spirit as the construction of nearly independent bonds for silicon with which we began this chapter. We obtain the corresponding bonding and antibonding orbitals in Appendix 5A as a function of the angle θ. We find that the energy is minimum with perpendicular orientation of the bonds, explaining why the bonding unit is bent, but bending it too much. We evaluate the energies at the observed 144° to obtain the energy-level diagram in Fig. 5.9.

We estimate the cohesion in Appendix 5A by subtracting the energy of the levels initially occupied (Si s^2p^2 as $2\varepsilon_s + 2\varepsilon_p$) from the energy of the final levels ($2\varepsilon_z + 2\varepsilon_x$), and add a $1/d^4$, fit to the equilibrium spacing. This gives -21.9 eV per bonding unit in comparison to the experimental -9.0 eV. Some of the discrepancy must come from Coulomb effects in the transfer energy, which we shall discuss for oxides in the next chapter, but probably the coupling is also overestimated. We successfully treated a wide variety of properties of SiO_2 in terms of essentially the same description of the electronic structure in Harrison (1999). Here we proceed to other systems.

Pantelides and Harrison did theoretical alchemy for silicon dioxide, which turned out to be quite interesting. We may transfer a proton between

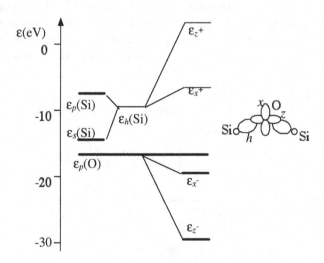

Fig. 5.9. Energy-level diagram for silicon dioxide, with the observed angle between bonds, formed from Si sp^3 hybrids and O p states, as sketched to the right. To the left, heavy lines represent occupied levels.

neighboring silicon nuclei to obtain $AlPO_4$, aluminum phosphate, another to obtain $MgSO_4$, magnesium sulphate, and a third to form $NaClO_4$, sodium perchlorate. $AlPO_4$ is quite insoluble in water, and has a structure analogous to SiO_2. $MgSO_4$ *is* soluble, and as Epsom salts occurs with seven water molecules per molecular unit imbedded in the structure. The Mg levels have much different energy from the oxygen levels so we could well think of it as separated Mg^{2+} ions and SO_4^{2-} ions. Although this is very much like SiO_2, a different bonding unit becomes appropriate. We again include only the p states on the oxygen, but we now take the tetrahedron centered on a sulphur as the bonding unit. We carried out the calculation of levels in Appendix 5B to obtain the energy-level diagram in Fig. 5.10. Note that the starting sulphur p state is above the oxygen levels, but the sulphur s state is below. It does not affect the final result qualitatively; either way we would obtain three p-like levels above one s-like level, all occupied, and another like set of four levels above it, all empty.

The starting energy levels *do* affect the cohesion, calculated in Appendix 5B. The energies of the states occupied in the atom are subtracted from that in the cluster states of the molecule for a -112.16 eV gain. The repulsion proportional to $1/d^4$ required to obtain the equilibrium spacing of 1.48 Å is found to be 43.78 eV, leading to a predicted cohesive energy of -68.38 eV, in comparison to the experimental $-28.$ eV (obtained also in Appendix 5B). The overestimate is slightly greater than that for SiO_2, which we again attribute to Coulomb effects for oxides discussed in Chapter 6, which are significant. The extension of the calculation to sodium perchlorate is

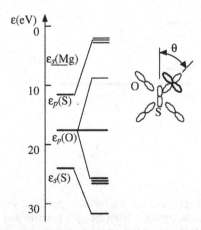

Fig. 5.10. Energy level for a sulphate ion from $MgSO_4$. One π-oriented p state from the oxygen tetrahedron is shown in heavy lines to the right.

immediate, and carried out in Appendix 5B, leading to a cohesion of −60.33 eV compared with the experimental −16.91 eV, a comparable overestimate.

The experimental values of cohesion — we compare Si_2O_4 with $MgSO_4$ and $NaClO_4$ — shows the same trend of decreasing cohesion for more polar systems that we found for the compound semiconductors, −38 eV for Si_2O_4, −28 eV for $MgSO_4$, and −17 eV for $NaClO_4$. Our predictions for the last two also show that trend, but not Si_2O_4. That discrepancy is not surprising; we used a different bonding unit for Si_2O_4 so the different approximation makes different errors.

CHAPTER 6

salt

Ionic Compounds

We think of the covalent solids in terms of bonds, often two-atom bonds but sometimes based upon larger bonding units. We think of ionic solids in quite different terms, as consisting of ions, with electrons transferred from one atom to the next. In Section 1.2 we indicated that the covalent semiconductors were understood doing theoretical alchemy on the diamond structure of the Column IV elements, but ionic solids by doing theoretical alchemy on the inert-gas atoms of Column VIII. For diatomic *molecules*, MgS for example, the distinction is not so sharp. The valence electrons are predominantly located on the sulphur so that we could think of it as an S^{2-} ion, but we could also think of it as a very polar Mg-S two-electron bond. In solids the distinction is quite clear; if a tetrahedral structure is formed we should make the approximate bond description of the solid. If it is in a more closely packed structure, like the rock-salt structure, favored by low electrostatic energy, we make approximations appropriate to the ionic description. MgS actually occurs in both structures, and we proceed differently in the two cases. It is characterized by the structure, not the composition. We begin with the simplest, purest case, the alkali halides.

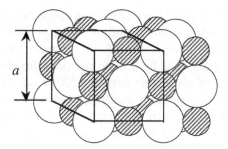

Fig. 6.1. The rock-salt structure. Large spheres (Cl⁻) lie at the corners of a cube and in the center of each face. Small spheres (Na⁺ or K⁺) lie at the center of each edge and at the center of the cube. In the periodic structure there are four ions of each type per cube of volume a^3. The nearest neighbor distance is $d = a/2$, 2.82 Å for NaCl.

6.1 Cohesive Energies

The theory of the cohesion of the alkali halides (e.g., Harrison, 1999) is extraordinarily simple. We imagine bringing the isolated neutral atoms together into the rock-salt structure shown in Fig. 6.1. We then remove an electron from the alkali atom, costing an energy $-\varepsilon_s = 4.96$ eV for sodium (from Table 1.1) and put it in a chlorine p state, at $\varepsilon_s = -13.78$ eV (same table), for a gain of -8.82 eV, reasonably close to the observed -6.8 eV per ion pair.

We have mentioned a number of times the question which arises immediately as to why it was not the electron affinity of the chlorine, the energy at which an electron is added to a neutral atom, which enters; it is higher by the Coulomb U for that atom, 10.3 eV for Cl from Table 1.1. The answer again is that there is a lowering of the energy for the electron on the Cl in the crystal due to the six neighboring positively charge Na ions, plus the reverse shift of the nearest negative Cl ions, etc. The sum of all of these charges out to large distances is a tricky problem. The result is called the Madelung energy for that structure, a potential ϕ at the negative ion site which shifts the electron energy at that site by

$$-e\phi = -\alpha Ze^2/d,\qquad(6.1)$$

with d the nearest-neighbor distance, 2.82 Å for NaCl, and α is called the *Madelung constant*, 1.75 for the rock-salt structure and smaller, 1.64, for the structure of the polar semiconductors which we discussed in Chapter 5. The

Coulomb interaction varying as $1/r$ is so long-range that even if we sum out to spheres large enough to contain millions of ions, it is still fluctuating in sign as we increase the sphere size. If one corrects for the fluctuation in *total* charge within the sphere by adding the opposite charge at the sphere surface, it converges to about one part in 1000 if one sums out to $300d$, and it is easy to write a program to do that (Harrison, 2006a). Much quicker convergence can be obtained by summing over finite crystallites if one is careful to keep the crystallites and their surfaces neutral (Baker and Baker, 2009. 2010). There is also a very long history of elegant techniques for doing these calculations to high accuracy, but when values are not available, these simpler calculations are frequently helpful. In the case of NaCl, the correction to the cohesion is $U - \alpha e^2/d = 1.36$ eV, bringing the first estimate of -8.82 eV to -7.46 eV, even closer to the experimental -6.8 eV. This agreement is representative of that obtained for all of the alkali halides (Harrison, 1999, p. 337). The experimental values generally are around -8 eV for the fluorides, dropping to about -5.5 eV for the iodides.

This concept of the cohesion is very much different from the traditional theory given by Born (1931). His was an empirical approach in which one again begins with isolated atoms, for example sodium and chlorine. One ionizes the sodium atom requiring experimentally an energy 5.14 eV. This electron is than added to a chlorine atom gaining an energy equal to the electron affinity, given experimentally by 3.6 eV, but which could be estimated from Table 1.1. One then assembles these ions to form a crystal, gaining an electrostatic energy equal to the Madelung energy of $-\alpha e^2/d = -8.9$ eV. Together these give us a cohesive energy of -7.4 eV, close to our estimate and to the experimental value. Born also added a small repulsive overlap interaction between atoms which he estimated.

This is simply a different path between the atomic and crystalline end points, but conceptually it attributes the cohesion to the electrostatic Madelung energy, where we attribute it to the transfer of electrons to atomic states of lower energy. Indeed, both approaches correspond to a cohesion of $-\varepsilon_s + (\varepsilon_p + U) - \alpha e^2/d$ (plus perhaps a repulsion), though he used experimental values rather than the theoretical term values from Table 1.1. Perhaps the most important difference is that Born's was a theory of cohesion, whereas ours is a theory of the electronic structure. We can in fact use ours, adding the couplings between orbitals on neighboring atoms, to calculate the entire range of properties of these systems, an effort undertaken in Harrison (1999). In addition, the Born theory makes it clear that when we make such corrections as the $-\alpha e^2/d$ that we are to use these ideal charges of $\pm e$ on the ions, rather than some effective charge Z^*, because in bringing the ions together almost all of the electrostatic energy is gained before the ions overlap each other and charges shift between them. Finally, Born's theory is not directly applicable to divalent compounds, those between elements of Columns II and VI, because the isolated atoms of Column VI will not accommodate a second electron. There is no difficulty in applying ours directly to these compounds.

The extension to divalent compounds is immediate and most of the compounds are even in the same rock-salt structure of Fig. 6.1. For CaS we bring the atoms together to their spacing of $d = 2.85$ Å, and then transfer two electrons to obtain $2(\varepsilon_p(S)-\varepsilon_s(Ca)) = -12.56$ eV, not far from the observed -9.7 eV cohesive energy per ion pair. To make the electrostatic correction, we note that when we add the first electron to S, we must add $U(S)$ for the electron affinity, and when we add the second, its final state is again up $U(S)$ as an electron affinity, but also another $U(S)$ from interaction with the first electron we transferred. Furthermore, taking the second electron from the Ca s state takes an extra energy $U(Ca)$ more than ε_s because of the absence of the first electron. Thus the correction is $3U(S) + U(Ca)$ and similarly, both electrons see a Madelung potential from doubly charged ions so that the correction to the cohesion is $3U(S) + U(Ca) - 4\alpha e^2/d = -0.62$ eV. It is too small to be significant, and in fact we would ordinarily discard a negative value as unrealistic. The agreement with experiment is similar, and the electrostatic corrections small, for all of the divalent compounds (Harrison, 1999, p.338) *except* for the oxides, for which the results are given in Table 6.1. There the corrections are considerable, and significantly improve the agreement with experiment. It is striking that the oxides are the problematic cases, and they are also the most important ones and the ones upon which we have focused.

Table 6.1. Cohesive energy for the monoxides, in eV per ion pair, predicted as $2[\varepsilon_p(X)-\varepsilon_s(M)]$ in the first entry. For the second entry $U(M) + 3U(X) - 4\alpha e^2/d$ was added. Experimental values, the third entry, were obtained in Harrison (1999).

	MgO	CaO	SrO	BaO
$2[\varepsilon_p(X)-\varepsilon_s(M)]$	−19.8	−22.9	−23.8	−25.0
Plus Coulomb	−17.1	−14.9	−13.8	−12.4
Experiment	−10.4	−11.0	−10.4	−10.3

It is good to know that the oxides are different, and to know how to make this relatively important correction for them. For the other compounds, the simplest theory of the cohesion is remarkably accurate.

6.2 Other Properties

We should note first what this theory suggests about the surface energy. We obtained the main contribution to the cohesion by transferring electrons between atoms, and we can do that just as well near the surface as in the bulk. Thus it suggests very tiny surface energies in comparison to the cohesive energy per bond broken, and this is in fact true. Particularly interesting is the contrast with covalent semiconductors, where the surface energy is very *large* in comparison to the cohesive energy per bond broken. This would be much more difficult to guess using Born theory. It is possible to calculate the extra electrostatic energy for forming a surface, and the methods of Baker and Baker (2009) are ideally suited to it. They indeed lead correctly to the small surface energies.

A second interesting property is the energy-band gap, which is very large in the ionic crystals, the reason that they are insulators rather than semiconductors. The first suggestion which comes to mind is a surprisingly good one: it might be the energy to transfer an electron *back* onto the metallic ion from which it came, $\varepsilon_s(M)-\varepsilon_p(X)$ for an alkali halide. This might look like a bad suggestion because we know that the energy levels are broadened into bands, as we saw for the covalent semiconductors in Chapter 5. However, a clue there turns out to be correct. We saw in Figs. 5.2, 5.3, and 5.4 that at the center of the Brillouin Zone ($k = 0$), the bands were either three-fold degenerate or nondegenerate, corresponding to p-like or s-like states. That is quite general for high-symmetry structures and is reflected in the different symmetry notations for the bands in Fig. 5.3. Correspondingly here, where the important coupling will be between the p states on the Column VII or VI atom and the s state on the neighboring Column IA or IIA atom, the effect of that coupling disappears at the center of the Zone and the simple theory will give the band energies at the free-atom energies. As k moves away from the center of the Zone, the coupling which arises will push the bands apart so that the difference between free-atom energies in not an unreasonable guess for the band gap. The bands at the center of the Zone in the covalent solids were also s-like and p-like, but the $V_{pp\sigma}$ and $V_{ss\sigma}$ for neighboring atoms was also important, shifting the bands at the center from the free-atom values. It is not even quite fair for the alkali halides since we shall introduce coupling between sodium s states and those on neighboring sodium ions in Section 6.3, but we proceed without that for the moment. For NaCl then we expect a band gap of $\varepsilon_s(Na)-\varepsilon_p(Cl)=8.8$ eV, close to the observed 8.5 eV and the agreement is similar (Harrison, 1999, p. 333) for all of the alkali halides, agreement comparable to that for the cohesive energy. The comparison for the divalent compounds was not quite as good, with again the biggest discrepancies for the oxides.

We must note that there are Coulomb corrections to the band gaps, even for full density-functional-theory calculations such as shown in Fig. 5.3. We mentioned them briefly for semiconductors in Section 5.4. They are easily understandable here, where we think of the excitation of an electron across the gap as the transfer to another atom. If the electron and the hole it left behind are far from each other, we should be adding a U to the energy required, with little cancellation from the e^2/r. However, in this dielectric medium the energy should be reduced by the dielectric constant of the crystal, just as it was for semiconductors, a correction U/ε. (This is not immediately obvious but can be seen to be reasonable. It arises from the polarization of the region surrounding this ion, lowering the potential at that ion.) As we noted in Chapter 5, this simple correction turns out to be rather accurate not only for these ionic crystals, but for covalent semiconductors and even inert-gas solids (Harrison, 1999, 206ff for more discussion of the gap enhancement). For silicon it gives an increase in gap by 0.65 eV, compared to the difference of 0.69 eV between density-functional-theory gaps and experimental gaps. These are small on the scale of the inaccuracies of the tight-binding bands. For rock salt they are large, enhancing the gap by $U/\varepsilon=3.5$ eV. The couplings between sodium s states which we mentioned in the last paragraph would reduce the gap by essentially the same amount. Thus it would be fair to say that we were lucky to find nearly the correct gap for NaCl. However, the success for the other alkali halides tells us that the near cancellation applies for all of these systems and allows us to use this extraordinarily simple picture for ionic crystals.

These predictions of the energy-band gap bring up another quite remarkable point. We have found a gap estimated as $\varepsilon_s(M)-\varepsilon_p(X)$, as well as a cohesion of $\varepsilon_p(X)-\varepsilon_p(M)$. Thus it suggests the empirical relation that the magnitude of the cohesion per ion pair is equal to the band gap for an alkali halide. Similarly, the magnitude of cohesion is predicted to be equal to twice the band gap for a divalent ionic solid. We are not aware that this relation between experimental quantities came up before it was raised by the tight-binding analysis. It is semiquantitatively correct since the individual relations were confirmed at least semiquantitatively.

Analysis of most dielectric and elastic properties requires the introduction of the coupling between states on adjacent ions, which can readily be done using the couplings we introduced in Chapter 3, or scaled values. Here we look only at the dielectric constant, related to the polarizabilies we have discussed for a number of systems. The first step is to define an effective charge Z^* for the ions. For rock salt the charge on the sodium consists of $+1e$ for the electron transferred to the chlorine, minus whatever electronic charge is transferred back by the interatomic coupling. The sodium s state is coupled to the σ-oriented p state on each of the six neighboring chlorine ions by $V_{sp\sigma} = (\pi/2)\hbar^2/(md^2)$ from Eq. (3.11). In perturbation theory each such coupling transfers $(V_{sp\sigma}/(\varepsilon_s-\varepsilon_p))^2$ electrons to the sodium. With six neighbors and two spins this leads to

$$Z^* = 1 - \frac{12V_{sp\sigma}^2}{(\varepsilon_s-\varepsilon_p)^2}, \qquad (6.2)$$

equal to 0.65 for NaCl with $d = 2.82$ Å. The chlorine ion has the negative of this. As we indicated, this Z^* should not be used for a Born-theory calculation of the cohesion, but may be appropriate for estimating potentials within, or near, the crystal. For divalent compounds, a 2 replaces the 1 in Eq. (6.2), but the second term is the same.

Of more interest is the movement of charges if a field is applied. A field $\mathbf{E}$ along an x direction will raise the energy ε_p of electrons on the nearest neighbor in the positive x direction by Eed transferring $V_{sp\sigma}^2 2eEd/(\varepsilon_s-\varepsilon_p)^3$ electrons (from the first derivative of the transfer) of each spin to the sodium. The same amount will be transferred to the chlorine in the minus-x direction so Z^* does not change but for each sodium this number of electrons is moved $2d$. Dividing the corresponding dipole by the volume $2d^3$ per sodium gives the polarization density, which we set equal to the susceptibility χ times the electric field to obtain a susceptibility of

$$\chi = \frac{4e^2V_{sp\sigma}^2}{(\varepsilon_s - \varepsilon_p)^3 d}. \qquad (6.3)$$

This does not include any displacement of the *ions* which the field might cause, so it is the susceptibility which determines the optical dielectric constant, $\varepsilon = 1 + 4\pi\chi$. For NaCl, Eq. (6.3) gives $\chi = 0.067$, compared to an experimental 0.10. A similar success is obtained for the other halides for sodium and lithium, but for the heavier alkali metals the measured susceptibility was much larger than predicted. When we included also excited p and d states on the alkali metals with the same approach (Harrison 2006b), we found that it accounted for the difference. Eq. (6.3) can be directly applied to the divalent compounds, CaS, etc., and is again found to be qualitatively correct, with largest discrepancies for the heavier metallic elements (Harrison, 1999). There also we made direct application to noble-metal halides and to ten-electron compounds such as PbS and TlCl. In the ten-electron compounds the metallic s states are occupied and the important coupling is between p states. Then the $V_{sp\sigma}^2$ in Eq. (6.3) is replaced by $V_{pp\sigma}^2 + 2V_{pp\pi}^2$ and the $\varepsilon_s(M) - \varepsilon_p(X)$ replaced by $\varepsilon_p(M) - \varepsilon_p(X)$.

It is interesting that this view of the dielectric susceptibility is quite different from the traditional theory of ionic solids, as was the view of cohesion. It has been customary to treat it as the sum of the polarizabilities of the individual ions, which might be calculated as we did for inert-gas atoms in Chapter 1. However, the approach was to adjust polarizabilities for the halogen atoms and alkali atoms to fit the observed susceptibilities of the compounds. With the systematic variation in the empirical values from compound to compound, there were enough parameters to fit them quite well. Similarly, the divalent compounds could be fit. However, with a *real* test, such as using these values for a cross compound such as CaF$_2$, it failed. The approach we have used here on the other hand is directly applicable to these cross compounds. We would prefer the approximate description without the many adjustable parameters, even if it were not very accurate. Actually for the heavier divalent compounds there was some indication (Harrison, 2006b) that it was appropriate to include coupling between orbitals on the *same* ion, as well as between ions, which would correspond to including polarizability of individual ions, as in the traditional theory.

Another interesting property is the effective charge associated with the dipole produced by displacing an ion, called the transverse charge, e_T^*. It contains the Z^* but also a contribution arising from the shortening of the front bond, giving extra charge transfer and the lengthening of the back bond, reducing the transfer. For alkali halides the result is

$$e_T^* = Z^* + \frac{16V_{sp\sigma}^2}{(\varepsilon_s - \varepsilon_p)^2} = Z + \frac{4V_{sp\sigma}^2}{(\varepsilon_s - \varepsilon_p)^2} \qquad (6.4)$$

with $Z = 1$, and with $Z = 2$ for the divalent compounds. While Z^* is always less than the formal charge Z, the transverse charge, is always greater. The transverse charge, in contrast to Z^*, is directly measurable from the splitting of the longitudinal and transverse optical-vibration frequencies. For NaCl Eq. (6.4) gives $e_T^* = 1.12$, compared to an experimental 1.11. This formula is applicable to all the rock-salt-structure compounds, requiring the generalization given above for the generalization to the ten-electron compounds, and the predicted effective charges are generally in much better agreement with experiment than the susceptibilities. It is an important point that what is appropriate for any effective charge depends strongly on just what property it is to be applied to. In the semiconductor compounds there is also a piezoelectric charge, which is quite different from the transverse charge (Harrison, 1999).

6.3 Color Centers and Molecular Ions

One feature which distinguishes the ionic compounds, with large energy gaps, from the semiconductor compounds is that the properties of holes can be very different from those of the electrons. An electron added to the empty conduction band in NaCl can move through the crystal, but the hole left behind "self-traps". This was discovered by Castner and Känzig (1957) in the mid-1950's using electron spin resonance to detect what is now called the V_k center. The hole, an empty chlorine p state, causes two adjacent chlorine ions to move together, emptying one of the antibonding $p\sigma$ states. It can move to a neighboring site through an electron excited from a neighboring Cl⁻ ion filling this state, with a new V_k center forming at the site from which the electron came. It is important to the understanding of the lack of conduction by holes in many ionic crystals. It is also is an example of a molecular ion forming in a solid.

With the discussion of the V_k center, we should perhaps mention also the more familiar F center in the alkali halides. Irradiation of alkali halides by x-rays, for example, colors them and the centers responsible were named after the F in *Farbe*, German for *color*. They are associated with vacant halogen sites, which would be formally of $+e$ charge, but capture an electron to become neutral. These have been traditionally thought of as an electron captured in the empty space, but in fact they should be described in terms of a tight-binding combination of the s states on the six neighboring alkali ions. [With no charge density in the vacant site, Poisson's Equation guarantees that the potential cannot be minimum.] The lowest state is a full-symmetry combination of alkali s states. Each s state is coupled by a $V_{ss\sigma}$ to four other s states in the cluster giving a cluster energy $\varepsilon_s + 4V_{ss\sigma}$ and the $V_{ss\sigma}$ describes second-neighbor interactions for separations of $\sqrt{2}d$ and a value -0.59 eV for NaCl with $d = 2.82$ Å if we use Eq. (3.8). [This is the coupling we mentioned in the preceding section which would reduce the gap by $6 \times V_{ss\sigma} = -3.54$ eV, but was cancelled by the Coulomb enhancement.] There is no second-order coupling through the chlorine p states because the $p\sigma$ states, to which each s state is coupled, are not coupled to any of the other Na s states in the cluster, in the absence of lattice distortion. The first excited states of this cluster of six Na ions will be odd in reflection through a (100) plane, with coefficients zero in the plane and of opposite sign on the other two ions. Then there is no coupling between contributing s orbitals and the energy of the odd cluster orbital is ε_s. We take the excitation energy for the electron bound to the vacancy then to be the $4|V_{ss\sigma}| = 2.36$ eV for NaCl. We may in fact take the expected excitation energy for the F centers which color all the alkali halides in various colors to be the same $(\pi^2/4)\hbar^2/md^2$ in terms of the alkali-halogen separation d. The two are plotted in Fig. 6.2 and the agreement is extraordinary. We would not have had confidence on this scale, either in the $1/d^2$ or in the coefficient. This may be the strongest evidence that our unscaled couplings are appropriate for alkali halides.

Following the discovery of the self-trapped hole, Känzig organized many studies of another example of such ions, the hyperoxides of sodium or potassium, NaO_2 and KO_2 (e.g., Rosenfeld, Ziegler and Känzig, 1978). Rather compete studies were completed and published in German before

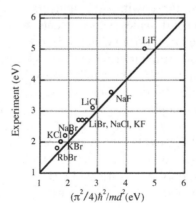

Fig. 6.2. Experimental excitation energies for F centers (Kittel (1976) p. 547), plotted against the elementary prediction.

they were noticed by others, leaving little incentive for other workers to continue the study. The O_2^- and Na^+ ions form in a rock-salt-like structure, with the O_2^- ions held together primarily by the $p\sigma$ bond. But one of the antibonding π states is also empty (two were empty in neutral O_2) so the molecular ion has a net spin of one electron. Thus these remarkable compounds have magnetic properties which occur ordinarily only in transition-metal and f-shell-metal systems such as we shall discuss in the next chapter. Perhaps the reason that they are so little known was Känzig's decision to complete their study before publicizing the results.

Crystals such as the $MgSO_4$ which we discussed in Section 5.6 as polar covalent insulators may also be described as based upon molecular ions. We showed the energy-level diagram in Fig. 5.10 for the covalently bonded SO_4^{2-} ion, though in that context we thought of it as covalently bonded to the Mg, and indeed the structure shows four Mg neighbors to each SO_4^{2-} ion as in the covalent semiconductors. Either view works, but if dissolved in water, the SO_4^{2-} ions remain intact, surrounded by water molecules with their protons close to the SO_4^{2-}. With hydronium ions, rather than Mn^{2+} ions, they provide sulphuric acid. We should consider these *oxyanions* separately from their role as contributors to these covalent insulators.

6.4 Dioxides

The production of acids such as sulphuric acid involves the solution of oxides into water. The dioxides (other than alkali hyperoxides) provide a good starting path toward the oxyanions. Most stable molecules have even numbers of electrons; those with odd numbers like CN are called *free radicals*. Thus we expect dioxides with elements in Columns II, IV, and VI. MgO is a typical ionic crystal from Column II, discussed in Section 6.1. We consider here SO_2 with sulphur from column VI and shall return to CO_2 with carbon from Column IV. The sulphur-dioxide molecule consists of a sulphur atom with two oxygen atoms a distance $d = 1.43$ Å away, with $119.5°$ between the two sulphur-oxygen separations, as illustrated in Fig. 6.3. This looks like the bonding unit we used for SiO_2, but now the oxygens are the outer atoms. In these oxides the s state on the central atom is not so deep and it is important to keep it, as well as the p states, as we shall see. For the oxygen atoms, only the σ-oriented p states are essential as indicated in the figure. In Appendix 6A we solve for the tight-binding electronic

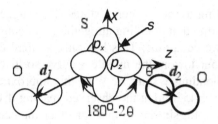

Fig. 6.3. An SO_2 molecule, central sulphur orbitals coupled to the σ-oriented p states on the oxygen atoms. $d_1 = d_2 = 1.43$ Å.

states based on these orbitals, as a function of the angle θ between the bonds. The resulting energies are reproduced here as Fig. 6.4. Adding the energies of the occupied states, and the repulsive energies from nonorthogonality, the minimum energy occurred at $\theta = 45°$, bonds normal to each other. They are correctly predicted to be bent, but at $90°$ rather than the observed $120°$ angle. Of particular interest were the curves themselves. The three labeled x are for states even in reflection through the vertical x axis in Fig. 6.3. The lowest of these can be identified with the sulphur s state at -24.02 eV, only slightly

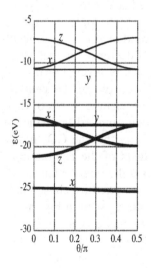

Fig. 6.4. Energy levels for SO_2 as a function of the angle θ. Occupied states are shown as heavy lines.

shifted by its coupling to the oxygen p states. This suggests simplifying the problem by eliminating that coupling altogether, reducing the problem to the solution of quadratic equations. This has only a small effect on the upper curves in Fig. 6.4 but lowers the middle x curve so that it meets the $\theta = 0$ axis where the π-bonding y state is. With this approximation, energy was minimum again at $\theta = 45°$ and at the observed spacing if we took a λ for the repulsions equal to 1.46, similar to that for other molecules. The energy at the minimum was -18.43 eV compared to the observed cohesive energy of -11.13 eV, agreement similar to that for similar molecules.

It was an interesting point that this approximation of dropping s states, as we did for the Column VI elements in Section 5.4, seemed the appropriate one, and correctly led to the bent structure (not quite the correct angle) for SO_2. When we form hybrid states in Appendix 6B, we find that the dangling hybrid caused this same bending. That dangling hybrid had the sulphur p-state energy of -11.60 eV at $\theta = 0$ and dropped to the sulphur s-state energy of -24.02 eV at $\theta = 45°$, crossing most of the energy range of interest in Fig. 6.4. We see in Appendix 6B that when the coupling with the s state is included, this dangling hybrid is so strongly mixed with the σ bonds and antibonds that one can scarcely see where it had been before coupling. Certainly for SO_2 the use of hybrid states would at first seem quite inappropriate, but we come back to this shortly.

We turn next to CO_2 which could be equally well represented by the sketch in Fig. 6.3, and would lead to energy levels very much like those in Fig. 6.4. The important difference is that there are two fewer valence electrons, so the highest-energy level shown occupied in Fig. 6.3 becomes empty. Had we dropped the role of the central-atom s state, so that the middle x curve matched the y curve at $\theta = 0$, that y curve would always be higher. It would be the one emptied and since it is independent of θ we would then predict incorrectly that the CO_2 molecule was bent just as SO_2 is. However, with the effects of the central-atom s state included, we see from Fig. 6.4 that at small θ the intermediate x state rises above this y state; it is the one that becomes emptied for those angles, lowering the energy at small θ. It was in fact found, as seen in Fig. 6A.2 in Appendix 6A, that this lowering was sufficient to make this $\theta = 0$ total energy the lowest and correctly predict a straight CO_2. Thus the inclusion of the central-atom s state is essential to understanding the difference in geometry between the SO_2 and CO_2.

We did not follow this up quantitatively with parameters for CO_2, but the most interesting lesson had been learned. The distinction between the geometry of CO_2 and SO_2 arises entirely from the effects of the central-atom s state, combined with the change in the number of electrons. In this qualitative sense an argument based upon hybrids, such as we made for semiconductor surfaces near the end of Section 5.1, is similar: it requires the use of the central-atom s state and if the dangling hybrid is occupied it is energetically favorable to bend the molecule to bring the dangling-hybrid energy down near the s state energy; the bend has a smaller effect in raising the bonds so the dangling hybrid dominates. If the dangling hybrid is empty, it is favorable to make the molecule straight to bring it up to the p-state energy, with some lowering of the energy of the bonds. This qualitative argument gives the correct geometry for SO_2 and CO_2 but one could worry that it might not always do so. It could have been that the effect of the s state was not large enough in CO_2, that the x curve rose so little above the y curve that the CO_2 ended up bent. However, it appears that hybrids *do* given the correct geometry in many cases and we return shortly to their application to the geometry of a wide variety of oxyanions, not forgetting that for anything at all quantitative about oxyanions, we must proceed in a more complete way.

This difficulty with the hybrids came as somewhat of a surprise because hybrids have proven so useful and effective in the semiconductors. We may look a little closer at the reasons this might be so. The question concerns the relative size of the covalent energy V_2 and the metallic energy V_1. This V_1 represents the effects of dehybridizing the states and should be small compared to V_2 if we hope to treat the bonding between hybrids through V_2 as the dominant effect. For an sp^n hybrid, V_1 can be seen to be $(\varepsilon_p - \varepsilon_s)/(n+1)$. In Eq. (5.3) we gave the form for the covalent energy for sp^3 hybrids, which is easily generalized, and regarding SO_2 as sp^2 sulphur hybrids and O_2 as sp hybrids, and using the scaling of $1/3$ for SO_2 and 0.205 from Table 3.1 for O_2, we obtain for the three systems the values in Table 6.2. It is clear that for diamond, and the other semiconductors (Si: $V_1 = 1.80$, and $V_2 = 5.36$ eV, respectively), the approach with hybrids makes very good sense. For the others it becomes very questionable. This is largely from the scale factors we have found here (V_2 would be 12.63 eV for SO_2 and 13.90 eV for O_2 without scaling), but also from the rapid growth in V_1.

Table 6.2. Metallic and covalent energies for three systems.

	V_1	V_2
Diamond	2.06 eV	12.40 eV
SO_2	4.14	4.64
O_2	8.63	2.49

Before moving on to solutions of these oxides in water, we should mention another class of dioxides which is important. They arise from the fourth column of the transition-metal series, Ti, Zr, and Hf, which we shall discuss in Chapters 7 and 8. Being the fourth in the series they can give up all of their valence electrons to the two oxygen constituents to form simple insulating crystals. They form in closely packed structures and are clearly ionic solids, rather than the covalent insulators such as silicon dioxide, which form more open structures. We could in fact arrive at TiO_2 with theoretical alchemy starting from inert gases $ArNe_2$. We transfer protons from each Ne to Ar to obtain CaF_2, fluorite, and another to obtain TiO_2. It is a simple ionic compound, though the lowest empty band is d-like, not s-like.

An interesting feature of these dioxide insulators is that if doped by replacing the four-valent Zr, for example, by trivalent yttrium the crystal maintains electrical neutrality by forming oxygen vacancies, positively charged defects since an O^{2-} has been removed, one oxygen missing for every two dopants. This is in contrast to semiconductors in which doping adds electrons to a conduction band or holes to a valence band. That requires too much energy in ZrO_2. Such vacancies can diffuse through the crystal, allowing it to serve as an *electrolyte* in a fuel cell. If a fuel such as methane is on one side of the crystal, it can remove oxygen (O^{2-}) from the crystal to form neutral CO_2 and $2H_2O$. In the process two electrons are given to a conducting anode at the surface. Then the resulting oxygen vacancies can diffuse to the opposite side of the crystal where air replaces the missing oxygen, taking two electrons from a conducting cathode to neutralize the positively charged vacancy. In this way the chemical energy obtained by oxidizing the methane is taken out directly as electrical energy. We turn next to a more general class of oxides.

6.5 Oxyanions and Solutions

Having an understanding of the different geometries of SO_2 and CO_2, we may move on to the general class of oxyanions. We shall do this using arguments based upon hybrids, which indeed did gave the correct geometry for these two molecules. We begin with the bonding units which contain four oxygen atoms — or ions, which arose in theoretical alchemy on silicon dioxide. They are usefully regarded as independent ions,

SiO_4^{4-} PO_4^{3-} SO_4^{2-} ClO_4^{1-}

silicate   phosphate   sulphate   perchlorate

They form an isoelectric series, all having the same orbital states occupied, but different numbers of protons in the central-atom nucleus. Similar oxyanions can be formed from the heavier elements, such as AsO_4^{3-} from arsenic, below phosphorus in the periodic table. They all form as regular tetrahedrons. In all cases, as we saw for sulphates in Fig. 5.10, all bonding orbitals are occupied, antibonding orbitals are empty, and any nonbonding oxygen p states are occupied. We would have come to the same conclusion if we had formed sp^3 hybrids on the central atom, formed two-electron bonds between individual neighbors, occupied all bonds and nonbonding oxygen p states and left all antibonding states empty, though we saw that there were quantitative problems if we took this too literally.

If we go to the *sulphite* ion, SO_3^{2-}, the electronic structure is quite different, because of three rather than four oxygens. We could again form molecular orbitals as for Fig. 5.10 for any chosen geometry, but here we choose hybrids. If the three oxygen ions formed an equilateral triangle with sulphur at the center we would form sp^2 hybrids on the sulphur and form a bond with each neighbor. The four neutral atoms were short a total of eight electrons of having full shells, so for the 2 − charge state we must have six empty states, the three antibonds we just formed. The three bonding states are filled, but so also is the sulphur π state, the p state oriented perpendicular to the bond. Thus, with hybrids, we expect the sulphur to move out of the plane, reforming the hybrids oriented toward the oxygen neighbors, and replacing the π state by a dangling hybrid perpendicular to the original plane of the molecular ion. Then energy will be gained as this doubly occupied dangling hybrid drops in energy toward the ε_s for sulphur, just as SO_2 bent to drop the energy of the dangling hybrid. We would predict bond angles of 90° but they are larger at 106°. In the corresponding oxyanions of selenium and tellurium in the same column the angles become closer to 90°, just as H_2S and H_2Se have angles closer to 90° than H_2O, simpler for heavier molecules.

We may expect an isoelectronic series such as we had for the anions with four oxygens,

PO_3^{3-} SO_3^{2-} ClO_3^{1-}

(phosphite?)   sulphite   chlorate

and in fact chlorate has the same structure with nearly the same angle, but PO_3^{3-} does not seem to occur. The "phosphite" name has been taken over by HPO_3^{2-} with the PO_3 having similar pyramidal geometry with H^+ in the dangling hybrid.

It is interesting to ask what happens if we remove two electrons from SO_3^{2-}. The answer is that we have neutral sulphur trioxide, and we empty the highest-energy occupied states of the sulphite, the dangling hybrids. Thus, as for CO_2, the energy will be lowest if the empty dangling hybrid is at the sulphur p-state energy and the molecule should remain flat, as it does. This molecule is part of another interesting isoelectronic series (except that the central atom for all but sulphur comes from the next row up in the periodic table),

$$BO_3^{3-} \qquad CO_3^{2-} \qquad NO_3^{1-} \qquad SO_3^{0}$$

orthoborate	carbonate	nitrate	sulphur trioxide

and they are all planar as is sulphur trioxide. The isoelectronic oxyanions from the sulphur row of the periodic table seem not to arise. Neither does the ClO_3^{1+} which would come to the right because such positive trioxides do not occur.

This brings us back finally to the dioxides. Starting with SO_2 there seem not to be any PO_2^{1-}, etc., just as there were none from the same row in the SO_3^{0} series above. However, from the next row up we might start with OO_2, written O_3 and construct a series. (We could have obtained the series above starting with O_4, but that molecule seems not to occur.)

$$NO_2^{1-} \qquad OO_2^{0}$$

nitrite	ozone

As in SO_2^{0} the central-atom π states are empty, as well as the two antibonds, but the dangling hybrid is doubly occupied. The molecule, or molecular ion, is bent. CO_2^{2-}, which would come next, does not occur. We can, however move to the right of SO_2 if we add two electrons in the central-atom π states, giving ClO_2^{1-}, chlorite. The dangling hybrid is still doubly occupied so the molecular ion should be bent, as it is.

We could in fact add *one* electron to SO_2 to obtain SO_2^{-} which apparently exists (Gillespie, 1972), but it, like the ClO_2^{0} with which it is isoelectronic are free radicals, which we have not discussed here. We would expect the extra electron in chlorine dioxide to be in the π state (e.g., see

Fig. 6.4), with the dangling hybrid still doubly occupied, so we would again expect it to be bent, as it is at some 117°.

We see that the geometry of this very wide variety of oxyanions is correctly understood in terms of the simple hybrid picture. We expect that each could be understood in more detail, to the degree that we were able to understand the properties of SO_2, with simple tight-binding theory such as that for Fig. 6.3. We have learned that this should not be done with hybrids, and even the more promising simplification of neglecting the central-atom s state may not be adequate, as it was not for CO_2. However, the calculations are still quite simple even if the conceptual model is not quite as simple. We might note also that the usual chemical terminology given below each oxyanion above does not correlate very well with the electronic structure, represented by the various isoelectronic series of systems. Of course that is understandable since the terminology considerably predates the understanding of the electronic structures. We obtained much of the structural information about these molecules and molecular ions from Gillespie (1972), but he had a very much different view of the origin of the particular structures.

We return finally to dissolving a molecule such as SO_2 in water, with water viewed as in Chapter 2. SO_2 has no proton to transfer to a water molecule, as did HF when dissolved. However, there is the possibility of a water molecule giving up its two protons to other water molecules, forming two hydrondium ions, while the remaining O^{2-} attaches to the SO_2, making SO_3^{2-}, a sulphite ion. This separation of charge could not occur without the other water molecules orienting themselves to lower their electrostatic energy, and thereby *screening* the charged ions, as occurred for other solutions described in Chapter 2. This sulphite will have the pyramidal geometry which we just found. It is not so easy to convert a sulphite ion, SO_3^{2-}, to a sulphate ion, SO_4^{2-}, since that requires adding a neutral oxygen, breaking the bond to another oxygen, or to an H_2, which it previously had. However, if the SO_2 were dissolved in hydrogen peroxide, H_2O_2, it could acquire an O_2^{2-} to become a sulphate ion, and that is one way sulphuric acid can be made. In the same way, if a CO_2 were dissolved in water and picked up an O^{2-}, with two fewer electrons than SO_3^{2-}, it leads to a planar carbonate ion, CO_3^{2-}.

iron

CHAPTER 7

Transition and f-Shell Metals

The world would have been much simpler if only the s and p states which we have been discussing had been important. However, like many features of nature, if it had been that different life would not have been possible and we would not have been here to study it.

7.1 Transition Metals

The transition metals form a block, shown in Table 7.1, which can fit just to the left of Table 1.1, ending with Cu, Ag, and Au. [The first row, with copper, are called $3d$ metals, because in the hydrogen atom the lowest energy d state has the same energy as the $3s$ state, just as the lowest p state in hydrogen has the same energy as the $2s$ state and is called $2p$.] As we go from atom to atom entering the first transition-metal series from Table 1.1 from K with one s electron, Ca with two s electrons, we come to Sc with two s electrons and a d electron. In subsequent steps we ordinarily add one d electron at each step, until we doubly occupy all five d states, with two electrons each, at Cu. [There are some steps, such as Ni to Cu, where we add two d electrons, and drop to one s electron, mostly not of importance here.] The d state energy ε_d listed in Table 7.1 is lower at each step, but the energy

Table 7.1. Hartree-Fock term values for transition-metal elements from Mann (1967). All are for the $d^{X-2}s^2$ configuration (except D11, which are $d^{X-1.5}s^{1.5}$). Also given are the Coulomb U_d and the d-state radius r_d from Harrison (1999). The r_d values are from fits to Muffin-Tin-Orbital theory, which we now regard as preferable to the Atomic-Surface Method values, also listed there. They lead to the band widths, W_d, given as the fifth entry.

	D3	D4	D5	D6	D7	D8	D9	D10	D11
	Sc	Ti	V	Cr	Mn	Fe	Co	Ni	Cu
ε_s eV	−5.71	−6.04	−6.33	−6.59	−6.84	−7.08	−7.31	−7.54	−7.72
ε_d eV	−9.35	−11.05	−12.54	−13.88	−15.27	−16.54	−17.77	−18.97	−20.26
U_d eV	5.3	5.4	5.3	5.1	5.6	5.9	6.3	6.5	6.9
r_d Å	1.332	1.159	1.060	0.963	0.925	0.864	0.814	0.767	0.721
W_d(eV)	5.27	6.26	6.83	6.75	5.78	4.88	4.38	3.80	2.83
	Y	Zr	Nb	Mo	Tc	Ru	Rd	Pd	Ag
	−5.33	−5.67	−5.95	−6.19	−6.39	−6.58	−6.75	−6.91	−7.05
	−6.80	−8.46	−9.98	−11.49	−13.08	−14.61	−16.16	−17.70	−19.23
	1.9	2.4	2.7	3.0	3.4	3.7	4.0	4.3	4.6
	1.696	1.515	1.370	1.285	1.197	1.127	1.066	1.012	0.960
	6.76	8.43	10.02	9.99	9.52	8.49	6.95	5.56	3.67
	Lu	Hf	Ta	W	Re	Os	Ir	Pt	Au
	−5.42	−5.71	−5.97	−6.19	−6.38	−6.56	−6.71	−6.85	−6.98
	−6.63	−8.14	−9.57	−10.97	−12.35	−13.74	−15.12	−16.47	−17.78
	2.0	2.8	3.0	3.1	3.2	3.4	3.5	3.5	3.6
	1.693	1.553	1.433	1.361	1.284	1.219	1.159	1.116	1.081
	7.84	9.62	11.46	11.50	11.36	10.39	8.93	7.22	5.41

at which an electron would be added to a d state, the electron affinity, is higher than ε_d by a U_d, also listed in Table 7.1. This U_d is different from the U listed for the sp elements of Table 1.1 in that it yields the level to which an electron would be transferred from an s state on the same atom, rather than brought from far away. Thus it is essentially the repulsion between two d electron on the same atom *minus* the repulsion between a d electron and an s electron on the same atom. In that sense it is a *screened* Coulomb repulsion; in the crystalline metal this d-s repulsion would be replaced by a redistribution of metallic electronic charge.

When we form a metallic crystal from the transition element, the s states form a free-electron band just as they did in the simple metals in Chapter 4. The atomic d electrons have orbitals much closer to the nucleus, because of their two units of angular momentum, as we illustrated in Appendix 2B, so their coupling with neighboring state is weak and they form narrow bands, within the broad s bands, as we shall see in Fig. 7.3. We look first at the coupling producing those bands.

We had shown one d state in Fig. 1.1 which was a spherically symmetric function times zx/r^2. [The usual representation of angular-momentum eigenstates has the form of spherical harmonics, $Y_2^m(\theta,\phi)$, but we are taking a combination of them, called *cubic harmonics*, convenient for cubic symmetry.] There are two others which look the same, but oriented in different perpendicular planes, xy/r^2 and yz/r^2, shown in Fig. 7.1. If the atom is placed in an environment with a potential having cubic, rather than full spherical, symmetry, these three have the same energy, and are called t_g states. There are two other states, called e_g states which have the same energy as each other, one $(x^2-y^2)/r^2$ which is the same as an xy/r^2 state rotated $45°$ around the z axis, and another is $(3z^2-r^2)/r^2$, all shown in Fig. 7.1.

There are couplings between neighboring d states analogous to those we have given for coupling between p and s states. If the angular momentum is quantized around the internuclear axis, only states of the same angular momentum around that axis are coupled to each other, labeled by σ, π, and δ, for 0, 1, and 2 quanta of angular momentum. These also describe the coupling between cubic harmonics as illustrated in Fig. 7.2. Values for different relative orientation of these states, analogous to what we showed for p states misaligned from the internuclear axis at the bottom of Fig. 3.1, are obtainable from the Slater-Koster Tables (Slater and Koster, 1954), also given in Harrison (1999), p. 546.

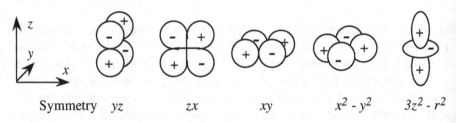

Symmetry yz zx xy $x^2 - y^2$ $3z^2 - r^2$

Fig. 7.1. Schematic sketch of the d-state cubic harmonics.

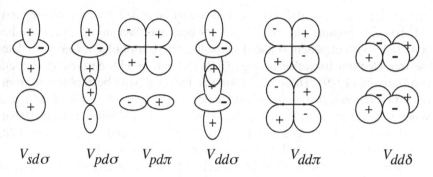

$$V_{sd\sigma} \qquad V_{pd\sigma} \qquad V_{pd\pi} \qquad V_{dd\sigma} \qquad V_{dd\pi} \qquad V_{dd\delta}$$

Fig. 7.2. The definition of matrix elements in terms of cubic harmonics. All here are negative except $V_{pd\pi}$ and $V_{dd\pi}$.

Approximate values for these couplings are obtainable as (Harrison, 1999),

$$V_{ddm} = \eta_{ddm} \frac{\hbar^2 r_d^3}{md^5} \qquad (7.1)$$

in terms of the r_d given in Table 7.1, and with

$$\eta_{dd\sigma} = -\frac{45}{\pi}$$

$$\eta_{dd\pi} = \frac{30}{\pi} \qquad (7.2)$$

$$\eta_{dd\delta} = -\frac{15}{2\pi}.$$

The ratios 6:–4:1 are the same as obtained in Muffin-Tin-Orbital theory (Andersen, Klose, and Nohl, 1978), as is the proportionality to $1/d^5$. They are derived in Harrison (1999) by evaluating the coupling between these atomic states and plane waves orthogonalized to them (OPW's given by $|k>> = |k> - |d><d|k>$). The coupling between two d states $|d>$ and $|d'>$ was then obtained as a second-order coupling, summing over intermediate states, $<d'|H|d> = \sum_k <d'|H|k>><<k|H|d>/(\varepsilon_d - \varepsilon_k)$. Similar expressions for coupling between f states, with three units of angular momentum, are $V_{ffm} = \eta_{ffm}\hbar^2 r_f^5/md^7$, the variation with spacing being in general $1/d^{2l+1}$. Applying such an approach to s and p states gives variation as $1/d$ for s-s coupling and $1/d^3$ for p-p coupling but we use $1/d^2$ for both and then any r_i appears in the numerator with exponent 0 and disappears. These formulae actually predate the free-electron formulae of Eqs. (3-8) through (3-11), derived in Froyen and Harrison (1979). These $1/d^2$ formulae had originally been obtained from fits to semiconductor energy-band calculations, e.g., $V_{ss\sigma} = -1.40\hbar^2/md^2$. When such formulae as Eq. (7.1) could be *derived* for d states, it seemed that a derivation for the s and p states must be possible, and a free-electron fit came immediately to mind. 1.40 was replaced by $9\pi^2/64 = 1.39$.

There is a simple method for calculating the r_d for any element, called the Atomic Surface Method (Straub and Harrison, 1985). It came from a conscious attempt to match two very different approaches to finding band widths. For the first, Andersen (1973) noted that Wigner and Seitz had found the bottom of a band by constructing a sphere of volume equal to the atomic volume in a solid, centered on an atom, and seeking an atomic-like solution with the slope $\partial R(r)/\partial r = 0$ at the sphere radius r_0. Andersen then sought the top of the band the same way, with the value $R(r_0) = 0$. Straub then sought to identify this with linear combinations of atomic orbitals, even midway between atoms for the bottom of the band and odd for the top, giving $R(r_0)$ equal to twice the value for the atomic state and $\partial R(r)/\partial r|_{ro}$ equal to twice the atomic value. This related the band width to the *atomic* wavefunction at r_0. It led in the end to a band width given by

$$W = \frac{-4\pi r_0^2 \hbar^2}{m} \frac{\partial \rho(r)}{\partial r}\bigg|_{r_0} \tag{7.3}$$

with $\rho(\mathbf{r})$ being the spherical average of the $\psi^*(\mathbf{r})\psi(\mathbf{r})$ for the atomic state upon which the band is based (normalized within the sphere). We will be writing the band width also in terms of these matrix elements, Eqs. (7.1) and (7.2) proportional to r_d^3 and can therefore equate the expressions to obtain r_d entirely in terms of the free-atom states. These turned out to be quite close to values obtained by fitting band widths given by Muffin-Tin Orbital (MTO) calculations of the band width (Andersen and Jepsen, 1977), and both are listed in Harrison (1999). It turned out subsequently (Harrison, 2007) that the MTO values give slightly better results and we have used them here. There is little incentive to use the simpler Eq. (7.3) when the values in Table 7.1 may do better.

In any case we now have the couplings needed to make an elementary band calculation, such as those for lithium in Appendix 4A, or for semiconductors in Chapter 5. Froyen (Harrison and Froyen, 1980) gave formulae for the bands in terms of these parameters for transition metals in face-centered-cubic, body-centered-cubic, and hexagonal-close-packed, structures (see Appendix 4B for structures). For wavenumbers in a [100] direction in a face-centered-cubic structure, the d bands were

$$\varepsilon_k(xy) = \varepsilon_d + 3V_{dd\sigma} + V_{dd\delta} + 4(V_{dd\pi} + V_{dd\delta})\cos(^1\!/_2\,ka),$$

$$\varepsilon_k(yz) = \varepsilon_k(zx) = \varepsilon_d + 2V_{dd\pi} + 2V_{dd\delta} + (3V_{dd\sigma} + 2V_{dd\pi} + 3V_{dd\delta})\cos(^1\!/_2\,ka),$$

(7.4)

$$\varepsilon_k(x^2 - y^2) = \varepsilon_d + 4V_{dd\pi} + (^3\!/_2\,V_{dd\sigma} + 2V_{dd\pi} + {}^3\!/_2\,V_{dd\delta})\cos(^1\!/_2\,ka),$$

and

$$\varepsilon_k(3z^2\text{-}r^2) = \varepsilon_d + V_{dd\sigma} + 3V_{dd\delta} + {}^1\!/_2(V_{dd\sigma} + 12V_{dd\pi} + 3V_{dd\delta})\cos(^1\!/_2\,ka),$$

(with the first corrected from Harrison (1999); in the top line the second $V_{dd\delta}$ was given as $V_{dd\sigma}$ and the ε_d was omitted). He also added a free-electron band and incorporated a coupling $<k|H|d>$ which came from the theory described after Eq. (7.2). Froyen plotted the bands for all the transition metals (ibid) in their correct structures, but the trends were obscured by the

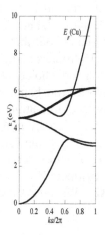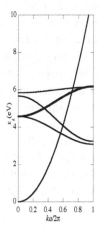

Fig. 7.3. The energy bands of fcc copper, on the left with hybridization between the d and free-electron bands and on the right without that hybridization, both from Eqs. (7.4). The heavy line is a doubly degenerate band. a is the cube edge; $d = a/\sqrt{2}$.

changes in structure. Here we plot the face-centered-cubic (fcc) bands, with parameter chosen for copper, on the left in Fig.7.3, showing how the d bands are inserted in the free-electron band. Copper is at the end of the $3d$ series with the d band completely filled and the bands are filled to the Fermi energy E_F indicated in the figure to accommodate one electron per atom in the free-electron band. As we move left in the series, Ni, Co, etc., we see from the bottom line in Table 7.1 that the band width W_d steadily increases. Also, with one less electron the d bands must rise above the Fermi energy leaving empty states at the top of the d bands. If all of these metals were in the fcc structure the trends across each series are easy to imagine though with many bands at the Fermi energy the Fermi surface could be complex.

7.2 The Friedel Model and Cohesion

It is helpful to return to the d bands shown to the right in Fig. 7.3, before the introduction of hybridization with the free-electron bands. They contain states for exactly ten electrons per atom, with a distribution in energy $n_d(\varepsilon)$ which is complicated in detail but which can be approximated by a constant distribution,

$$n_d(\varepsilon) \approx 10/W_d \text{ for } \varepsilon_d - W_d/2 < \varepsilon < \varepsilon_d + W_d/2, \text{ and } 0 \text{ otherwise}, \qquad (7.5)$$

the Friedel Model (Friedel, 1969) which we introduced in Section 5.5 for graphite and illustrated in Fig. 7.4. The total density of states is the sum of the free-electron contribution, $n_{fe}(\varepsilon)$, and the d-band contribution. $n_d(\varepsilon)$. It provides a greatly simplified description of the electronic structure.

We found (Harrison, 1999, 559ff) that a good choice of the Fermi energy put 1.5 electrons per atom in the free-electron band, with the Z_d other electrons occupying the d bands, near the free-atom occupation. Transferring any electrons in or out of the d bands rapidly moves ε_d up or down to keep Z_d nearly the same, because the U_d in Table 7.1 is so large. This immediately fixes E_F in Fig. 7.4 as $\hbar^2 k_F^2/2m$ with k_F chosen such that the volume of the Fermi sphere is ¾ that of the Brillouin Zone, or in terms of Ω_0 (the atomic volume), $4\pi k_F^3/3 = ¾ (2\pi)^3/\Omega_0$. We *could* take W_d from

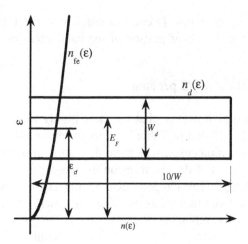

Fig. 7.4. The Friedel Model of the electronic structure of a transition metal.

Eq. (7.3), but we shall determine it here by fitting the second moment of the bands as we did for graphite. Then the position of the d bands, ε_d, is adjusted to give the correct occupation of these bands at the previously determined Fermi energy E_F relative to the free-electron-band minimum. (Thus in this case the ε_d in Table 7.1 based on $Z_d = 2$ becomes irrelevant.)

The second moment of the d bands is calculated from the tight-binding $5N$-by-$5N$ Hamiltonian matrix (N atoms with 5 d orbitals each) from which the bands are calculated, $M_2 = (1/5N)\sum_{I,j} H_{ji} H_{ij}$. The diagonal elements are all ε_d and do not contribute so for X nearest neighbors to each atom we obtain the second moment

$$M_2 = X(V_{dd\sigma}{}^2 + 2V_{dd\pi}{}^2 + 2V_{dd\delta}{}^2)/5, \tag{7.6}$$

which we equate to the second moment of the square distribution $W_d{}^2/12$ to obtain

$$W_d = \sqrt{12X(V_{dd\sigma}^2 + 2V_{dd\pi}^2 + 2V_{dd\delta}^2)/5} = 8.93\sqrt{12X}\,\frac{\hbar^2 r_d^3}{md^5} \tag{7.7}$$

using Eqs. (7.2). Note that $\sqrt{(12X)} = 12$ for fcc crystals.

Given these parameters for the Friedel Model we can immediately estimate the contribution of the d bands to the cohesion. We start with W_d equal to zero and all d electrons at the same ε_d. When we broaden the band, the Z_d electrons per atom (the D-column number at the top of Table 7.1 minus 1.5) occupying the lower portion of the band gives us an estimate of the energy gain as

$$E_{coh} \approx -(1 - Z_d/10)Z_d W_d/2 \text{ per atom}. \tag{7.8}$$

We must not forget, however, that for these highly localized orbitals the Coulomb repulsion U_d listed in Table 7.1 can be important. It has the same effect as for diatomic molecules, which we described in Section 3.6, in that it produces a segregation of the electrons on the two atoms, and weakens the effect of the coupling. This happens also in the crystal and in Harrison (1999, p. 599) we showed that it has the effect of generalizing Eq. (3.23) for diatomic molecules by replacing the W_d in the energy of Eq. (7.8) by $\sqrt{(W_d^2 + U_d^2)} - U_d$ for crystals. We should also remember that in all cases where we introduced an attraction between the atoms, there needed to be an additional repulsion which, in the simple metals, cancelled half of the energy gain. We apply that here and the contribution to the cohesive energy per atom from Eq. (7.8) becomes

$$E_{coh} = -\frac{Z_d}{4}\left(1 - \frac{Z_d}{10}\right)\left(\sqrt{U_d^2 + W_d^2} - U_d\right) \tag{7.9}$$

relative to free atoms. We evaluated this for each of the transition metals using the parameters of Table 7.1 as an estimate of the effect of the d bands on the cohesion. If the metal occurred in a structure different from fcc, we used a d for an fcc structure at the same density. The resulting E_{coh} was very small for the noble metals (−0.13 eV per atom for Cu, −0.30 eV for Ag, and −0.69 eV for Au) so they may be regarded as simple metals. Thus in Fig. 7.5 for each metal we subtracted the value of Eq. (7.9) for the noble metal at the end of that row, to compare with the experimental cohesive energy of each metal, with the experimental value for the noble metal subtracted.

The theory inevitably does not reproduce the irregularities shown by the experiment, but the overall agreement is quite remarkable. The partial filling of the d bands is clearly the origin of the increase in cohesion at the center of the series, as proposed by Friedel (1969), noting the $Z_d(10 - Z_d)$ dependence of the energy. It is responsible as well as for the accompanying reduction in interatomic spacing, and increased hardness. It is also interesting that this extra cohesive energy is not so structure-dependent, and so the high cohesion does not carry over to a high melting point. Further, we

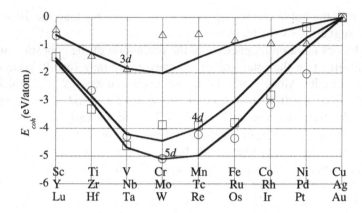

Fig. 7.5. The cohesive energy of each transition metal, relative to that of the noble metal at the end of the series. Lines are from Eq. (7.9); points are from experiment[a], identified at an element where they agree.
[a]Kittel (1976).

may see what it suggests for surface energies. It is proportional to the square root of the number of neighbors to each atom, so if a surface atom loses three of its twelve nearest neighbors, its contribution to the energy is reduced by a factor 0.87 rather than the 0.75 which would be estimated by bond breaking, as for the simple metals not so far off.

The only important choice in this derivation was the 1.5 free electrons per atom which came from noting that the sum over states in full band calculations, weighted by the d-like contribution, tends to lead to this value, even in copper. Wills and Harrison (1983, described also in Harrison (1999), 563ff) showed that quite remarkably this choice of 1.5 free electrons includes approximately the effects of sd hybridization. Even in the noble metals, one effect of hybridization is to add some s-like (or free-electron-like) character to the occupied d bands, increasing the s-like occupation by some δZ_s of order 0.5 per atom to the effective s-occupation, which could be estimated in perturbation theory. They similarly showed that this lowered the energy by an amount that could be calculated in perturbation theory (based upon the coupling between the d states at low energy and the empty free-electron states at high energy). This lowering turned out to be just what one would estimate by adding a $\delta Z_d = -\delta Z_s$ to the Z_d in Eq. (7.8) or (7.9).

Thus even in the noble metals we use a Z_d of 9.5, rather than the 10 which would be suggested by the band structure.

We did not mention magnetism in the discussion of the cohesion, and it is a small effect in the total energy. However, we may readily address it using the Friedel Model. We need the exchange energy U_x which is obtainable from the free-atom spectra as we did for s and p states and is listed for the transition metals in Table 7.2. Then we can allow the ε_d to differ for up and down spin, keeping the number of d electrons fixed by adjusting the same Fermi energy for both. This raises the band energy of Eq. (7.8) (calculated now separately for up and for down spin) but lowers the U_x exchange energy and we find that the exchange energy wins if $U_x > W_d/5$ (Harrison, 1999, p. 589). If we incorporate the effects of U_d as in Eq. (7.9), the criterion becomes $U_x > W_d^2/[5\sqrt{(U_d^2 + W_d^2)}]$. Values are compared in Table 7.2. The only three for which the criterion is satisfied are Fe, Co, and Ni, which are in fact the only magnetic transition metals. Iron could have gone either way; it is ferromagnetic in the bcc structure, but nonmagnetic in the fcc structure. Co and Ni have more complicated magnetic structures.

We note that we have treated the volume-dependent energy of the d electrons quite separately from the free-electron contributions. The energy

Table 7.2. Values for U_x in eV obtained (Harrison, 1999, p. 591) from atomic spectra (Moore (1949, 1952)) in eV, and values for $W_d/[5\sqrt{(U_d^2 + W_d^2)}]$ in eV obtained from Table 7.1. Systems with $U_x > W_d/[5\sqrt{(U_d^2 + W_d^2)}]$ are predicted to be ferromagnetic.

Z	4	5	6	7	8	9	10
3d Series	Ti	V	Cr	Mn	Fe	Co	Ni
U_x	0.90	0.68	0.64	0.78	0.76	1.02	1.60
$W_d/[5\sqrt{(U_d^2 + W_d^2)}]$	0.95	1.08	1.08	0.83	0.62	0.50	0.38
4d Series	Zr	Nb	Mo	Tc	Ru	Rh	Pd
U_x	0.63	0.48	0.60	-	0.88	0.81	-
$W_d/[5\sqrt{(U_d^2 + W_d^2)}]$	1.62	1.94	1.91	1.79	1.56	1.20	0.88
5d Series	Hf	Ta	W	Re	Os	Ir	Pt
U_x	0.70	0.60	0.39	0.36	0.61	0.86	-
$W_d/[5\sqrt{(U_d^2 + W_d^2)}]$	1.85	2.22	2.22	2.19	1.97	1.66	1.30

as a function of spacing for each contribution can be thought of as six times the contribution of each nearest-neighbor interatomic interaction. In Harrison (1999) we plotted these two contributions separately along with the sum, as shown in Fig. 7.6. We used such curves to treat the elastic properties and structure determination, but here we move on to f-shell metals, where such curves will be particularly useful.

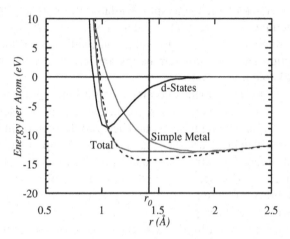

Fig. 7.6. The d-state and free-electron contributions to the energy of Cr, plotted separately, and with the total, as a function of spacing (more precisely, r is the radius of a sphere with the atomic volume). The dotted curve includes adjustments to bring the minimum to the observed value.

7.3 Rare-Earth Metals

The first places where f states are occupied in the free atoms are the rare earths, with $4f$ electron, and fit to the right of Ba in the lower part of Table 1.1. The actinides have $5f$ electrons and appear to the right of Ra just below that. The f orbitals are even more localized than the d electrons, for the same reason, and in the case of rare earths have almost no interaction with f states on neighboring atoms. Then we can think of them as purely atomic states within the metal. Such f states have almost no effect on the chemical properties, and it is therefore difficult to separate one rare earth from another in the laboratory. In that sense the f electrons have almost no effect on the cohesion of the rare earths, and yet the cohesive energy of the series is very interesting, as described in Harrison (1999, 614ff).

After the divalent barium, comes the trivalent lanthanum (neither with f electrons) and much larger cohesive energy, as a trivalent metal, than the divalent barium, as we saw for such simple metals in Chapter 4. Next is cerium metal with one f electron in this similar trivalent (not counting f electrons) metal. Then comes praseodymium with two f electrons and again trivalent, etc. The energy is lowest if the f electrons on each atom have parallel spin, due to the exchange energy U_x, and following Hund's Rule. With $2l + 1 = 7$ states of the same spin available the magnetic moments on each atom become very large. They interact very weakly with each other through their interaction with the free-electrons, called the Ruderman-Kittel (1954) interaction and for some rare earths are ferromagnetically aligned to produce remarkable magnets.

Actually, the trivalent metals continue only to europium, which should have six f electrons but takes on a seventh, completing the seven-electron shell of parallel spin, leaving it divalent. Then gadolinium again has seven f electrons and is trivalent, as are subsequent rare earths up to ytterbium which should have thirteen f electrons but takes on a fourteenth to complete the shell and is divalent. This seems to be the basis for a false myth that there is some extra stability in having a full shell, presumed to be from correlation energy. We would really expect the correlation energy per electron to be less, if any different, for full shells because there are no empty levels to utilize in correlating electron motions.

We may see the flaw in the argument by looking at the cohesion. Our discussion would suggest that the trivalent metals should have the same strong cohesion, but that the divalent Ba, Eu, and Yb should have the same weak cohesion. We see in Fig. 7.7 that indeed Ba, Eu, and Yb have weak cohesion and that the La, Gd, and Lu which follow them have strong cohesion, but they actually vary rather linearly with number of electrons Z between La and Eu and between Gd and Yb.

The discrepancy with our expectation does not have anything to do with bonding in the metal, but in properties of the atom. Of all of these elements, only La, Ce, Gd, and Lu are trivalent as atoms, $s^2 d(f^{Z-3})$ (Weast, 1975, B3), and these form trivalent metals with the expected large cohesion; the rest are divalent as atoms. Among these Ba, Eu, and Yb form divalent metals with the expected small cohesion. For the others one electron must be promoted from a d state to an f state. The d-state energy is about the same through the series, but starting with Ce with one f electron, at each step the energy of the f level should drop by U_x from exchange with an additional f electron, so the promotion energy and the magnitude of the cohesion should drop by U_x

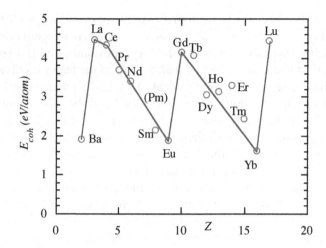

Fig. 7.7. The magnitude of the cohesive energy of the rare earths, from Kittel (1976). The line interpolates linearly the promotion required from divalent atoms to trivalent metals (from Harrison, 1999).

at each step, as they do approximately, seen by the straight line in Fig. 7.7. The extrapolation of that line from Sm to Eu could be taken as a prediction of the energy of *trivalent* Eu, though the point is divalent Eu. This suggests that the promotion energy in Eu is zero and there is no extra stability for the full shell.

The same description can be applied to the filling of the shell of opposite spin in Gd through Yb, and the entire spectrum of cohesive energies in Fig. 7.7 is understandable as arising from divalent and trivalent simple metals, plus the promotion energies arising from a U_x of about 0.5 eV.

This same myth about the extra stability of closed shells has been used to explain *valence-skipping compounds* such as compounds of lead, in which lead appears as divalent or tetravalent, but never as trivalent (Harrison, 2006c). In fact the difference in ε_s and ε_p in the heavy elements is extra large because of relativistic effects (heavy mass of s electrons when near the nucleus), not reflected in Table 1.1 which was based on nonrelativistic calculations but those tables suffice here. Then lead can form a divalent ionic compound PbS because its two occupied p states ($\varepsilon_p = -6.53$ eV) are shallower than the empty sulphur p state ($\varepsilon_p = -11.60$ eV), but it cannot form a tetravalent PbS_2 because its occupied s states are deeper ($\varepsilon_s = -12.49$ eV). It can however form PbO_2 because oxygen's p states are deeper ($\varepsilon_p = -16.77$ eV). Oxidizing lead can give either of these compounds, and if a compound Pb_2O_3 forms, it is accomplished by making half of the lead ions divalent and the other half tetravalent, none trivalent. This has been attributed to extra stability in having one atom with a full s shell and the other empty, rather than one s electron in each. The electrons effectively attracted each other as if the net repulsion U between them were negative. In fact, this negative-U behavior only occurs due to the displacement of neighboring atoms, different for the two valences, inward for the tetravalent Pb and outward for the divalent Pb. The system is analogous to, and the mathematics the same as, weights on springs as illustrated in Fig. 7.8. If we start with one weight on each spring, and the springs relaxed to minimize the energy, we can change to both weights on one spring, and again let both springs relax. We find the total energy is lowered by $-U_{neg}$ equal to half the gravitational energy gain, as if the two weights attracted each other. Similarly two heavy balls on a mattress will tend to roll towards each other. In the solid such a gain must exceed the extra Coulomb repulsion between the two electrons on the same site if it is to become a negative-U system. If the solid has a high dielectric constant ε, it reduces the Coulomb repulsion by a factor of $1/\varepsilon$, but never changes its sign.

Cerium is a special case among the rare earths. The room temperature phase we have described above has a single localized f electron on each atom. At high pressure or at low temperature it makes a phase change, with a 10% reduction in volume, though both phases may be face-centered-cubic. Certainly in the low-volume phase the f electron must become band-like, presumably in an f-like band, contributing to the cohesion and the attraction between atoms. The behavior is reminiscent of the actinides and the interactions which will be represented in Fig. 7.9.

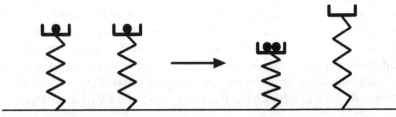

Fig. 7.8. A model of a negative-U center. One weight on each spring has higher energy than both on one, with both springs relaxed. The weights effectively attract each other.

7.4 Actinides

The actinides are the row of atoms under the rare earths to the right of radium, with partly occupied $5f$ shells. With the actinides we must consider the coupling between neighboring atoms, and they are analogous to those given for the d states in Eqs. (7.1) and (7.2). For example, $V_{ff\sigma} = (5250/\pi)\hbar^2 r_f^5/md^7$, with expressions for the others, and values for r_f given in Harrison (1999). The resulting band widths are given in Table 7.3, along with the Coulomb repulsion U_f, the counterpart of U_d for the transition metals. When applied to the rare earths, they lead to vales of W_f much less than U_f, corresponding to our description of the f electrons as localized on individual atoms. For the actinides the two are comparable, as seen in Table 7.3, indicating that we may expect band-like behavior for the lighter actinides. We see there that the spacing decreases as we proceed down the series, as in the transition-metals series, and we would correspondingly expect the cohesive energy to grow in proportion to $Z_f(1 - Z_f/14)$, though it is seen in Table 7.3 to be erratic. Beyond plutonium the spacing dramatically increases, suggesting rare-earth-like behavior, and localized electrons, consistent with the $U_f \gg W_f$. Indeed plutonium exists in many phases. We have given values for band-like α-plutonium; there exists also a δ phase with atomic sphere radius $r_0 = 1.81$ Å, face-centered cubic and rare-earth-like.

Table 7.3. Parameters for the actinides from Harrison (1999), including configuration in the atom and atomic-sphere radius r_0 in the metal.

	Z	Atom[a]	r_0 (Å)	W_f (eV)	U_f (eV)	E_{coh} (eV)[b]
Ac	3	ds^2	2.10	12.9	3.00	−4.25
Th	4	d^2s^2	1.99	5.56	3.20	−6.20
Pa	5	f^2ds^2	1.80	3.79	3.35	
U	6	f^3ds^2	1.69	3.76	4.09	−5.55
Np	7	f^4s^2	1.66	3.09	3.90	
α-Pu	8	f^6s^2	1.67	2.50	4.61	−3.60
Am	9	$f^{77}s^2$	1.91	0.98	4.96	−2.73
Cm	10	f^7ds^2	2.03	0.82	5.10	−3.86

[a]Weast (1975), B4, [b]Kittel (1976).

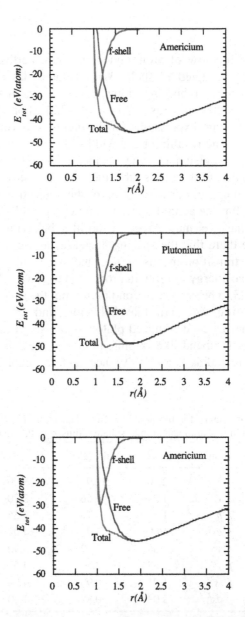

Fig. 7.9. The total energy of neptunium, plutonium, and americium as a function of atomic-sphere radius (from Harrison, 1999). Also shown separately are the f-shell and free-electron contributions.

The cohesive energies are complicated by the irregularities in the atomic configurations, indicated in Table 7.3, as they were for the rare earths. The cohesion, and many other properties are discussed in Harrison (1999, as well as 2004a and 2004b). Of particular interest is the energy as a function of atomic sphere radius, such as we gave for chromium in Fig. 7.6. These curves for Np, Pu, and Am from Harrison (1999) are given in Fig. 7.9. They seem to clearly characterize the transition from band-like to rare-earth-like behavior at plutonium. f states in neptunium are analogous to d states in chromium and other transition metals in that the effect of the f states combines with the free-electron states to produce a smooth curve with the minimum shifted to lower energy and deepened. In americium, the minimum in the f-state contribution to the energy occurs at such small spacings that the equilibrium spacing is determined almost entirely by the free-electron terms and the f shells have almost no impact on the bonding, as we found for the rare earths. In plutonium, the two contributions are close to providing two minima, one determined by the f-bonding and one by the free-electron bonding, which we of course associate with alpha and delta phases, respectively. The delta phase is fcc, as are many simple metals and rare earths, while the alpha phase has a complex structure, as does the transition metal manganese. It is interesting that this complex α-Pu structure is reminiscent of the kinds of $90°$ angle structures of the Column VI elements, S, Se, and Te, as we described in Section 5.4. Indeed the antisymmetric f states can favor such bonding just as the p states did for Column VI (Harrison, 1999, 624ff).

The complexity of the phase diagram for plutonium has led to a variety of speculations about electron correlations, just as the changing of valence between two and three for the rare earths did. Clearly our view is that the properties of plutonium are quite consistent with the relatively simple behavior of rare earths on the one hand and transition metals and light actinides on the other.

rust

CHAPTER 8

Transition-Metal Compounds

This is an extraordinarily large topic, and a very important one. We shall focus on some aspects that seem to us most interesting and relevant, and relate them to the other compounds we have discussed in the preceding chapters.

8.1 Electronic Structure

The transition-metal constituent will be described in terms of their s states and their d states, with energies given in Table 7.1. The nonmetallic atoms entering the compound are described by the levels given in Table 1.1. For the coupling between the metal s states and the p states of the nonmetals entering the compounds we use the same $V_{sp\sigma} = (\pi/2)\hbar^2/md^2$ from Eq. (3.11) which we have used for other systems. In Chapter 7 we gave coupling between d states, and the same formulation leads (Harrison, 1999) to

$$V_{pd\sigma} = -\frac{3\sqrt{15}}{2\pi} \frac{\hbar^2 \sqrt{r_d^3 r_p}}{md^4} \tag{8.1}$$

and

$$V_{pd\pi} = \frac{3\sqrt{5}}{2\pi} \frac{\hbar^2 \sqrt{r_d^3 r_p}}{md^4},$$

with r_d for the transition-metal atom given in Table 7.1 and r_p for the nonmetallic atom given in Table 8.1. As for the other systems we have treated we may include the nonorthogonality of neighboring orbitals by shifting their energies due to any coupling by $\lambda V_{ij}^2 / \sqrt{(\varepsilon_i \varepsilon_j)}$.

We shall initially look at compounds in the rock-salt structure shown in Fig. 6.1, taking chromium nitride as an illustration of those for which an energy-band description is appropriate. We return afterward to MnO. FeO, CoO, and NiO for which localized d states are a better starting point. CrN has a spacing of $d = 2.11$ Å, which leads to a $V_{pd\sigma} = -1.545$ eV and $V_{pd\pi} = 0.892$ eV. Then the tight-binding band calculation is simple, as for the transition-metal bands in Chapter 7. The σ band based upon the d state of $3z^2 - r^2$ symmetry for $\mathbf{k}$ in a [001] direction is

$$\varepsilon_k = \frac{\varepsilon_d + \varepsilon_p}{2} \pm \sqrt{\left(\frac{\varepsilon_d - \varepsilon_p}{2}\right)^2 + 4V_{pd\sigma}^2 \sin^2 kd}. \tag{8.2}$$

It happens that $\varepsilon_d(\mathrm{Cr})$ is almost identical to $\varepsilon_p(\mathrm{N})$ and we have raised ε_d by 0.5 eV to better illustrate the bands. The result is shown as light lines in Fig. 8.1(a). Also shown are the corresponding π-bands based on d states of zx, and yz symmetry (so doubly degenerate), the same as Eq. (8.2) but with $V_{pd\sigma}$

Table 8.1. r_p values, in Å, for atoms of nonmetallic elements, to be used to obtain matrix elements in Eq. (8-1). (After Harrison and Straub (1987).) ε_p values in eV from Table 1-1 are also listed for convenience.

	r_p	$-\varepsilon_p$		r_p	$-\varepsilon_p$		r_p	$-\varepsilon_p$
C	6.59	11.07	N	5.29	13.84	O	4.41	16.77
Si	13.7	7.59	P	11.4	9.54	S	10.1	11.60
Ge	14.4	7.33	As	13.2	8.98	Se	12.1	10.68
Sn	18.0	6.76	Sb	16.8	8.14	Te	15.9	9.54
Pb	19.8	6.53	Bi	18.9	7.79	Po	17.9	9.05

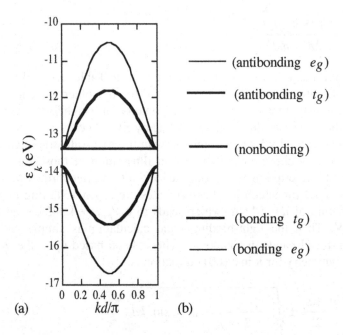

Fig. 8.1. In Part (a) are the CrN pd bands, obtained with Eq. (8.2) with the Cr d state raised from Table 7.1 by 0.5 eV. Doubly-degenerate bands are drawn as heavy lines. In Part (b) these bands are replaced by cluster levels, as in a special-points representation.

replaced by $V_{pd\pi} = -V_{pd\sigma}/\sqrt{3}$ and two nonbonding δ-bands based on states of symmetry x^2-y^2 and xy for which there is no pd coupling in Eq. (8.2) and the bands are flat, nonbonding. These bands in fact have the same energy throughout the Brillouin Zone; for three degenerate states coupled to five degenerate states, two of the five-set will be uncoupled, forming nonbonding bands. This was also true in ordinary rock salt for the three p states coupled to one s state, giving two nonbonding p bands, though we did not mention that explicitly in Chapter 6.

Chromium contributes six electrons (d^4s^2) and nitrogen three electrons (s^2p^3 but the s electrons occupy the deep nitrogen s states, not the pd bands), so the three bonding bands are each filled with one electron of each spin and there are three electrons left to fill three quarters of the states in the flat nonbonding band. Those bands will have some width if a more complete calculation is made, so the system is metallic, but with high resistivity because the bands are so narrow.

We could calculate total energies by constructing a Friedel Model, but it should be based upon square densities of states stretching above ε_d and below ε_p and it would be a little complicated. In a system with each band either full or empty, it is more convenient to use the *special points* method (e.g., Harrison, 1999, 348ff and described here as Appendix 8A) and this will have other useful features. In this method the average of the energy over the band is replaced by the value at a special wavenumber k^*. If we were really interested in a one-dimensional bands, perhaps approximately $2V_{ss\sigma}\cos(kd)$, we would pick $k^* = \pi/(2d)$ so that this dominant term of a Fourier expansion of the band was zero. In three dimensions the bands could be obtained in the same form as Eq. (8.2), but with $4V_{pd\sigma}^2 \sin^2 kd$ replaced by a squared sum over neighbors of couplings with appropriate $\mathbf{k}$-dependent phase factors. We see in Appendix 8A that by expanding the result for weak coupling, and identifying that result with the perturbation-theory result, we can obtain the band energies at the special point as Eq. (8.2) with $4V_{pd\sigma}^2 \sin^2 k^* d$ for one band replaced by a $V_2^2 = 6V_{pd\sigma}^2$ and for two other bands by a $V_2^2 = 6V_{pd\pi}^2$. The other two bands are nonbonding. The six in these expressions came from the six neighbors to each oxygen. It seemed quite remarkable to be able to find the result without ever deciding what wavenumber $\mathbf{k}^*$ was, nor getting the complete form for energy bands in three dimensions. Then the average energy of each antibonding or bonding band is taken to be

$$<\varepsilon> = \frac{\varepsilon_d(Cr) + \varepsilon_p(N)}{2} \pm \sqrt{V_2^2 + V_3^2} \qquad (8.3)$$

with either of these V_2's and with V_3 equal to the appropriate $(\varepsilon_d(\mathrm{Cr}) - \varepsilon_p(\mathrm{N}))/2$, near zero in this case. These levels are shown in Part b of Fig. 8.1.

8.2 AB Compounds with Bands

Perhaps the simplest transition-metal compounds can be obtained with theoretical alchemy as we did for other ionic compounds. For the compounds of interest we begin with a rock-salt structure of neon and argon, and move protons from the neon to the argon, making successively KF,

CaO, ScN, and TiC, moving into the transition-metal series which comes after Ca, but which we placed to the left of Cu in Table 1.1. These are basically simple ionic compounds, with just enough electrons to fill the bonding bands in Fig. 8.1, but the band gaps decrease at each step as the p states to the left of Ne rise and the d states of the metal drop. ScN is a semiconductor with a gap we estimate as 4.5 eV and, with $\varepsilon_d(\text{Ti}) \approx \varepsilon_p(\text{C})$, we correctly expect TiC to be conducting. In KF and CaO the lowest-energy empty bands are s bands, but at Sc the d state is lower than the s state and the empty conduction band is d-like. Thus both ScN and TiC must be regarded as transition-metal compounds [as was TiO_2 in Section 6.4]. Presumably empty d bands are also significant in CaO, but we did not include them in our discussion of such materials in Chapter 6.

The cohesion of ScN and TiC is estimated as with other ionic compounds. The largest part of the cohesion comes from transferring the two s electrons from shallow states to the nitrogen p stated for −16.26 eV for ScN and one d electron for −4.31 eV for a total of −20.57 eV. For TiC the s-electron transfer gains only −10.06 eV and the d-electron transfer makes negligible change. There seems not to be an experimental value for ScN but that for TiC is −14.6 eV, not so far from our −10.1 eV estimate. We can estimate the additional contribution of the interatomic coupling to the cohesive energy of TiC. With the p and d states of almost the same energy, we estimate a gain of magnitude $V_2^{eg} = 3.59$ eV for each spin in the e_g band for the observed $d = 2.16$ Å, and four times $V_2^{tg} = 2.39$ eV for the two spins in the two t_g bands, but half of that will be cancelled by the repulsion proportional to $1/d^8$ which will be there. Thus the correction is −8.37 eV making it slightly too large (−18.4 eV) in comparison to experiment (−14.6 eV).

Both of these compounds are called *octet* compounds because they have eight valence electrons, counting the oxygen s electrons, just enough to fill all the bonding states in Fig. 8.1. A consideration of a wide range of transition-metal monoxides, mononitrides, and monocarbides, generally in the rock-salt structure (Harrison and Straub, 1987) indicated that there are many with nine, ten, eleven, or twelve valence electrons, with the additional electrons entering the nonbonding band. However, with only a few exceptions, none with more than twelve, presumably because it requires inserting electrons in *antibonding* bands, and the system chooses other compositions. We regard the d electrons in the few exceptions (MnO, FeO, CoO, and NiO) as localized. Thus the description of the bonding in all of the other AB compounds in terms of simple energy bands seems to be reasonably sound. We turn next to the exceptions.

8.3 Localized States

For the $3d$ oxides of Mn, Fe, Co and Ni we start over again, following Harrison (2008). The isolated manganese atom has a configuration d^5s^2, and the five d electrons have their spin aligned, for a moment of 2.5 Bohr magnetons. We take the ε_d of Table 7.1 to be an average of the energies and using the exchange energy U_x of Table 7.2 we distinguish majority-spin and minority-spin levels as $\varepsilon_d{}^{maj}(\text{Mn}) = -17.22$ eV, and $\varepsilon_d{}^{min}(\text{Mn}) = -14.10$ eV. All majority-spin, and no minority-spin d states are occupied. Actually $\varepsilon_d{}^{min}(\text{Mn})$ represents the energy to which a majority-spin electron would move if its spin were flipped; an electron would be transferred from the s state on the same atom, or from a neighboring atom, to a d state at $\varepsilon_d{}^{min}* = \varepsilon_d{}^{min} + U_d = -8.50$ eV with the U_d from Table 7.2. These are localized states in the atom and we think of them as localized states in the oxide.

MnO forms in the rock-salt structure with $d = 2.22$ Å. In forming MnO we would transfer two electrons from the $\varepsilon_s = -6.84$ eV of Mn to the oxygen p states, leading to a contribution of -19.86 eV per Mn for the cohesion. If we correct it for the Coulomb terms as we did for the simple divalent compounds in Chapter 6 this is reduced to -13.70 eV, not so far from the observed -9.5 eV, with the five d electrons and their moment remaining on each Mn ion. We return at the end of this section to the super-exchange coupling between these moments.

In Mn_2O_3 a third electron might be transferred to an oxygen p state from a majority-spin d state from each Mn, leaving it Mn^{3+}, but the two states have almost the same energy. We must include the effects of pd coupling for this localized d state at the same time. We do this for MnO_6 clusters of the same geometry as that in rock salt (as in Harrison, 2008). As for molecular orbitals there is again a covalent energy V_2 representing the coupling between each central-atom d state and the combination of oxygen p states of the same symmetry. That squared coupling, $V_2{}^2$, is equal to the sum of the squared coupling with each neighbor. The majority-spin e_g state of the form $3z^2 - r^2$ is coupled to oxygen p states on two neighbors along the z axis by $V_{pd\sigma}$ and to states on four lateral neighbors by $-V_{dp\sigma}/2$ (using the Slater-Koster (1954) Tables), leading to a $V_2 = \sqrt{3}V_{pd\sigma}$, and the same result is obtained for the other e_g orbital of symmetry $x^2 - y^2$. Similarly, each of the three t_g orbitals leads to a $V_2 = 2V_{pd\pi}$. For the majority-spin d states the corresponding $V_3 = (\varepsilon_d^{maj} - \varepsilon_p)/2$ and we may obtain the energy of the *cluster orbital* from the counterpart of Eq. (8.3) (written there for Cr and N). Similarly, for minority-spin d states we would have the same V_2's but a $V_3 = (\varepsilon^{min}* - \varepsilon_p)/2$.

It is interesting that these V_2's are different from those we obtained for special points in the energy bands in the preceding section, but that is not a problem since we now think in terms of localized electronic states. It will turn out in the perovskite structure, with the same cluster geometry, that the special point in the energy bands would lead to the same V_2's that we obtain here for the clusters.

To illustrate the resulting states we have used the spacing for the perovskite $LaMnO_3$, rather than for MnO, to obtain the levels shown in Fig. 8.2, analogous to Fig. 8.1(b) obtained before introducing the exchange splitting. In MnO all majority-spin states are occupied. In Mn_2O_3 one upper majority-spin e_g electron is transferred to oxygen ε_p, and in MnO_2 a second transferred for an equal contribution. Minority-spin shifts and repulsions were added to obtain the total pd effects in Table 8.2. We found (Harrison, 2008) that the Coulomb corrections, as made in Section 6.1, were not large in these compounds. The same calculations can be made for the other $3d$ oxides. An iron atom has the configuration $d^6 s^2$

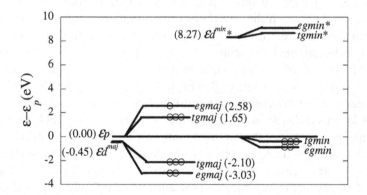

Fig. 8.2. The energy levels (in eV for $LaMnO_3$, measured from the oxygen p-state energy) for majority-spin (*maj*) and minority-spin (*min*) MnO_6 cluster orbitals. e_g states are doubly-degenerate, t_g states, triply degenerate. Circles indicated occupied states for Mn_2O_3 (or $LaMnO_3$). The upper *egmaj* level would have an additional electron in MnO, and no electrons in $SrMnO_3$.

Table 8.2. Contributions to the cohesive energy in eV per Mn, or Fe, from transferring manganese or iron s electrons to oxygen p states, from pd coupling and transfer, and the total (compared to experiment in parentheses).

		s to p transfer	pd effects	Total (experiment)
MnO	Mn^{2+}	-13.7	-1.37	-15.1 (-9.5)
Mn2O3	Mn^{3+}	-13.7	-3.04	-16.7 (-11.9)
MnO_2	Mn^{4+}	-13.7	-4.71	-18.4 (-13.5)
FeO	Fe^{2+}	-13.2	-1.42	-14.6 (-9.65)
Fe_2O_3	Fe^{3+}	-13.2	-1.85	-15.1 (-12.45)
FeO_2	Fe^{4+}	-13.2	$-3,20$	-16.4

with now a minority-spin state also occupied in FeO. The results of the corresponding calculations for cohesion are also shown in Table 8.2.

An important feature of these results is that the formal ionization states of the transition-metal ions, shown in Table 8.2, make integer changes from compound to compound. This arose also in Chapter 6 where lead was divalent Pb^{2+} in PbO, where there was a completely occupied s band, rather than tetravalent Pb^{4+} as in PbO_2. We also think of lead as tetravalent in the metal, where the bands only resemble free-electron bands if we include the s bands. It is still true in all cases that if we calculate the fractional occupation of the various atomic levels to obtain an effective charge Z^*, we will have intermediate values for all of these cases.

We return to the moments on the transition-metal ions, which are coupled to neighboring moments by Heisenberg exchange favoring their opposite alignment in an *antiferromagnetic* structure. The origin can be understood for two neighboring ions having opposite spin. Then an occupied majority up-spin state may be coupled to a minority up-spin state on a neighbor, lowering the energy of the occupied state. If the neighbor had parallel spin there would be two *occupied* states coupled and no energy gain. In the case of these magnetic oxides, the Heisenberg exchange is dominated by the direct V_{ddm} coupling between neighboring magnetic ions (Harrison, 2007), though these are more distant than the oxygen neighbors. There is a smaller contribution called *superexchange* (Anderson, 1950), which will be the dominant contribution to Heisenberg exchange when we consider the perovskites later in the chapter. For superexchange the coupling V_{ddm} is replaced by an indirect coupling through the neighboring oxygens and is obtained in second-order perturbation theory as $V_{pd\sigma}^2/(\varepsilon_d^{min*}-\varepsilon_p)$. Note that the energy denominator is quite large because of the ε_d^{min*} so perturbation theory is appropriate. The resulting coupling for two Mn ions on the opposite sides of an O is given by (Harrison, 2007)

$$\Delta E = -4 \frac{V_{pd\sigma}^4 + 2V_{pd\pi}^4}{(\varepsilon_d^{min*} - \varepsilon_p)^2 (\varepsilon_d^{min*} - \varepsilon_d^{maj})}. \tag{8.4}$$

[This is actually a factor of two larger than one might at first guess because it is a two-electron effect. We may already see the difference in the solution of the two-electron problem given in Eq. (3.23). If we expand it for small $V_{ss\sigma}$ we obtain a shift $-4V_{ss\sigma}^2/U^*$ rather than the $-2V_{ss\sigma}^2/U^*$ we would get in one-electron theory (Harrison (1999), 594ff).] A form similar to Eq. (8.4) applies for Mn neighbors at right angles to each other. These forms and values for the coupling were found to give a remarkably good account of the magnetic properties of these magnetic monoxides in Harrison (2007).

8.4 Perovskites with Bands

A particularly important class of transition-metal compounds is in the perovskite structure, illustrated in Fig. 8.3, distorted from this cubic arrangement for some compounds. A first step in understanding the electronic structure in such a system is counting the available electrons and states. For strontium titanate, $SrTiO_3$, called STO, shown in that figure, there are two s electrons from Sr, and four total of s and d electrons from Ti in the column D4 of Table 7.1. The three O atoms can accommodate a total of six extra electrons. Thus we correctly expect an insulator as long as the Ti d states are above the O p states, as they are. The s states on both Sr and Ti will be very shallow and not central to the electronic structure. The

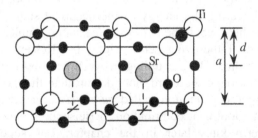

Fig. 8.3. The perovskite structure of $SrTiO_3$, STO, a simple cubic arrangement of titanium atoms (B sites), with an oxygen on every cube edge of length a, and with a strontium atom at each cube center (A site), for the formula ABO_3.

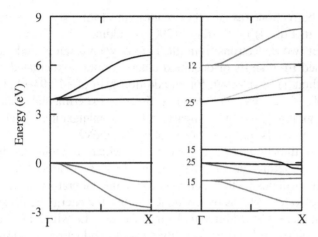

Fig. 8.4. The energy bands of $SrTiO_3$ for wavenumbers along a [100] direction. To the left are the bands obtained as in Eq. (8.2). To the right are the bands, adjusted to fit experiment, given by Mattheiss (1972). The differences can be accounted for by adding couplings $V_{sd\sigma}$ and second-neighbor pp-coupling.

lowest empty states will be predominantly Ti d states and we may expect energy bands to be formed (at least initially, since bands were formed in TiC, and in this case it turns out to be correct). They are very easily calculated as for Fig. 8.1, and are shown in Fig. 8.4.

The lowest conduction (empty) band, which is independent of energy in our bands to the left, would be based upon states of symmetry xy or x^2-y^2 for wavenumbers in a z direction. With two units of angular momentum around the z axis, they are uncoupled to p states, with zero or one unit of angular momentum, on the oxygens at $\pm d$ along the z axis. However, they are coupled by $V_{pd\pi}$ to p states on the oxygens in the x and y directions. Thus they are flat for wavenumber varying in the z direction, but rise quadratically in $\mathbf{k}$ variation in the x and y direction, giving rise to the middle curve if Fig. 8.4 were taken to represent bands for wavenumbers in the x direction.

The cohesive energy for $SrTiO_3$ is calculated in the same way as for the monoxides, but there are additional terms. We first transfer the two s electrons from the Sr and the two s electrons from the titanium to the oxygen p states for a gain or -45.28 eV, plus the Coulomb shifts in the ions and the

Madelung correction (as in the oxides in Table 6.1, generalized to perovskites in Eq. (4) of Harrison (2009)), adding +12.41 eV. Finally we must transfer two d electrons from the Ti to oxygen p states, and include the energy gained from shifts of occupied cluster orbitals, calculated as in Fig. 8.2, but without spin-splitting, for an addition gain of -13.94 eV for a total cohesion of -46.81 eV. We did not find an experimental cohesion for $SrTiO_3$ but judging from the comparison for the manganites which we shall discuss, it is probably an overestimate of some 10 eV.

Strontium titanate is a good system with which to begin any discussion of electronic structure for a perovskite, just as it is a widely used substrate upon which to deposit other perovskites in the laboratory. If we replace some of the divalent Sr ions by trivalent La ions, the energy bands are hardly affected, but there is an additional electron for each La which must appear in the d-like conduction band, making the crystal conducting, a doping just as in semiconductors. Replacing the Sr by monovalent ions similarly dopes the system with holes in the highest occupied, p-like bands, but there can be complications such as the self-trapped holes which we discussed for alkali halides. Replacing some of the titanium ions by ions from other columns of the transition metals will also dope the system, but also modifies the band structure, often describable by a weighted average of the two band structures (with additional scattering of any carriers by the difference in the two ion types). This possibility of controlling the properties as they are controlled in semiconductors has been responsible for a rapid growth in their study and applications in technology.

8.5 Manganites and Cluster Orbitals

It may not be surprising that when we replace titanium in STO by manganese in $SrMnO_3$, called SMO, the upper states in Fig. 8.4 may better be thought of as cluster orbitals. In fact we see from Fig. 8.3 that each Mn is again surrounded by an octahedron of oxygen ions as in the rock-salt structure. Thus the cluster orbitals are identical to those for the rock-salt structure and those we plotted in Fig. 8.2 were for a perovskite. Also as in the manganese oxides they are split into majority and minority spin levels, with some occupied and some empty. A wide range of properties of SMO was treated in Harrison (2009) and we discuss the most relevant ones here.

We mentioned in Section 8.2 that for perovskites the same energy would be obtained for special points in energy bands. In this case there are three oxygen p states for each of the three oxygens per formula unit, and only five d states, so we should match the special-point values to the d-state shifts. Then we obtain the same two e_g band averages and three t_g band averages that we obtained for the cluster states, as in Section 8.3, and four nonbonding oxygen p states.

This only slightly complicates an estimate of the cohesive energy in comparison to the rock-salt structure, and it was carried out in Harrison (2009). In $SrMnO_3$ it includes the transfer of the two s electrons from each Sr and each Mn to oxygen p states, exactly as for $SrTiO_3$. For these transfers it was important to include the Coulomb shifts of the levels as we did for the divalent simple compounds in Section 6.1. We included also the shift up of the minority-spin levels due to V_{pdm} coupling (with the corresponding lowering of some occupied oxygen states) which was a small effect, and the corresponding shift of the majority-spin states does not enter initially because both the bonding and antibonding states are occupied, but enters through the transfer of two electrons from the majority-spin e_g states to oxygen p states. Altogether this led to a cohesion for SMO of -35.46 eV, in comparison to the experimental -25.31 eV.

SMO is a rather simple insulating compound, though with the three upper majority-spin t_g states occupied it has a magnetic moment of three Bohr magnetons on each Mn cluster. These are coupled through Heisenberg exchange to produce an antiferromagnetic state, as were the oxides of Mn. In the case of the perovskite structure, there is almost no direct interaction between neighboring Mn ions and superexchange, as in Eq. (8.4), provides the coupling between moments, as described in Harrison (2009), and gives a good account of the magnetic properties.

The understanding of cohesion is similar for related compounds, such as $LaMnO_3$, called LMO, for which energy levels were given in Fig. 8.2. There is one more electron per formula unit, so only one manganese d electron is transferred to an oxygen p state, but one d electron is transferred from a shallow level in La, raising the estimate of cohesion to -41.49 eV, as the experimental cohesion is larger than for SMO at -30.42 eV. The cohesion varies linearly between the two compounds in $La_{1-x}Sr_xMnO_3$, called LSM, for the theory and also experimentally (Rømark, Stølen, Wiik, and Grande, 2002). However, the extra electron impacts other properties of this LSM compound strongly, making it a much more interesting system.

8.6   Conducting Manganites

These electrons in cluster orbitals do not flow through the crystal as they do in $SrTiO_3$, but it is possible for an electron to jump from one Mn to the next, decreasing the charge state of one and increasing that of the other, particularly if there were already different charge states as in a mixed crystal of $La_{1-x}Sr_xMnO_3$. Then a fraction x of the manganese ions are in a Mn^{4+} state and a fraction $1 - x$ in a Mn^{3+} charge state and the transfer simply interchanges a Mn^{3+} ion and a neighboring Mn^{4+} ion. Even then, an electron in a Mn^{3+} cluster will cause a relaxation of the neighboring ions, lowering its energy by $-W_p$, called a *polaron energy*. To move to a neighboring site where the relaxation has not occurred, it must coordinate its transfer with displacements of the ions, called polaron tunneling if it uses zero-point motion. Alternatively, it can jump to a higher-energy environment using thermal energy. This complicated behavior is called *small-polaron hopping* (e.g., Emin and Holstein, 1969). The conductivity arising from this small-polaron hopping is given by (ibid.),

$$\sigma = N(4e^2a^2/W_p)(\omega_0/2\pi)(W_p/2k_BT)\exp(-W_p/2k_BT), \qquad (8.5)$$

where ω_0 is a vibrational frequency characteristic of the lattice. These were plotted in Harrison (2009), and compared with experiment as in Fig. 8.5. It shows clearly the novelty of this conduction process. At low temperatures

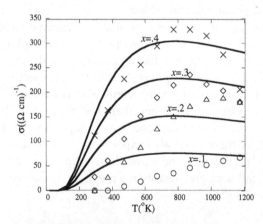

Fig. 8.5. Small polaron hopping conductivity. The lines are plots of Eq. (8.5). Data is from Tai, Nasrallah, Anderson, Sparlin, and Schlin, 1995, for the similar compound $La_xSr_{1-x}Co_{0.2}Fe_{0.8}O_3$.

where this hopping conductivity freezes out, there remains only the polaron tunneling, not included in the curves.

There is another consequence of these relaxations which occurs even in pure LMO. Then every manganese ion is Mn^{3+}, and either e_g state could be occupied on each. This single occupation of a degenerate state is exactly the condition for a Jahn-Teller distortion (e.g., Harrison 1999, p. 314), and Zhou and Goodenough (2002) have indicated that this occurs, an elongation of $\varepsilon = 0.08$ in a z direction and contraction of $-\varepsilon/2$ in the x and y directions. We independently estimated the size of the effect (Harrison, 2009) from the change in energy of the x^2-y^2 state linear in the distortion and the elastic energy from the experimental elastic constants, finding the energy minimum at $\varepsilon = 0.096$, close to the observed 0.08. The upper $3z^2-r^2$ states are lower than the x^2-y^2 states and are occupied, with only the upper x^2-y^2 states empty, again insulating, as observed. We also studied the magnetic states of this Jahn-Teller distorted state of LMO (ibid.) finding an antiferromagnetic coupling between neighbors in a z direction, but a ferromagnetic coupling to neighbors in the x and y directions. This ferromagnetic coupling arose from *double exchange* due to the coupling between the empty majority-spin x^2-y^2 state and the full majority-spin $3z^2-r^2$ state on a neighbor. These lead to the observed arrangement of (001) planes of parallel-spin ions, antiferromagnetically oriented with respect to neighboring planes in the z direction.

This magnetic ordering in LMO will disappear above some Néel temperature, T_N, as with other magnetic states. It was estimated (Harrison, 2009) as

$$k_B T_N \approx (S+1)\Delta E/S \tag{8.6}$$

in terms of the second-neighbor ΔE of Eq. (8.4), with $S = 2$ in $LaMnO_3$, and $S = 3/2$ in $SrMnO_3$. For $SrMnO_3$ this gives $T_N = 195°K$, comparable to the observed $260°K$. The estimate was complicated by the mixed magnetic structure of LSM and the estimate less successful.

At still higher temperature, $750°K$, the Jahn-Teller distortion will also disappear. Then the x^2-y^2 and $3z^2-r^2$ states are equal in energy and equally occupied. Electrons can move between neighboring sites without excitation and the resistivity drops greatly and becomes metallic, though with still a very high resistance in comparison to normal metals. In this state the magnetic moments remain, but with random orientations. The crystal is paramagnetic as expected. With such wide variations in resistance depending upon the magnetic state, it is not surprising that large changes in resistivity arise from magnetic fields, magnetoresistivity which has been widely studied. This was among the properties discussed in Harrison (2009).

Another distinction which many perovskites have from $SrTiO_3$ is that their surface planes, ordinarily (100) planes, are not neutral. In $SrTiO_3$ we may see from Fig. 8.3 that a surface plane contains one Ti^{4+} for each two O^{2-} ions and is neutral. So is the next, $Sr^{2+}O^{2-}$, plane. In $LaMnO_3$ the surface $Mn^{3+}O^{2-}{}_2$ planes have a charge of $-e$ per Mn, and the next plane a charge of $+e$ per La. It was noted long ago in connection with semiconductors (Harrison, 1979) that such charge arrangements lead inevitably to electric fields either inside the material or outside, and an arbitrarily large surface energy in bulk materials. In semiconductors these charges are ordinarily compensated for by surface reconstructions (Harrison, 1999, 734ff). A number of workers (Mannhart, Blank, Hwang, Millis, and Triscone, 2008) have found that at interfaces such as between $LaAlO_3/SrTiO_3$ this is compensated by electrons in the conduction band at the $SrTiO_3$ side of the surface, opening the door to new electronic devices. We might not expect this to occur in the manganites where we would expect the compensation to come by conversion of the charge state of a number of interface Mn ions, but these materials are generally conducting in any case.

Another interesting feature of the manganese perovskites, which we presume is also true for those of iron, cobalt, and nickel, is the small energy required to change the charge state of the clusters. Thus when La was substituted for some Sr ions in $SrMnO_3$, it simply converted some Mn^{4+} ions to Mn^{3+}. In ZrO_2 we noted in Section 6.4 that, in contrast, replacing Zr^{4+} by La^{3+} compensates the charge by forming vacant oxygen sites, because that takes less energy than placing electrons in the conduction band, high in energy.

This difference would seem to have an important consequence in LSM, particularly concerning its frequent role as a fuel-cell cathode. Vacancies in LSM again correspond to a missing O^{2-} ion, and we would expect charge neutrality to be achieved by replacing Mn^{4+} ions by Mn^{3+} ions, and the energy is lowest if they are adjacent to the vacant site. Thus an oxygen vacancy tends to form a neutral F-center, a missing oxygen ion with the two neighboring Mn sites having reduced charge states. In contrast, we noted that oxygen vacancies in an electrolyte such as ZrO_2 (Y-doped) are positively charged. If they diffuse into the cathode, the neutralization of the vacancies should occur naturally at the electrolyte-cathode interface, not at the cathode-air interface, as usually assumed (e.g., Lee, Kleis, Rossmeisl, and Morgan, 2009).

8.7 Oxygen at Surfaces

We recently (Harrison, 2010) made an application of the approach taken here to absorption of oxygen atoms and molecules at such surfaces. It may provide an interesting illustration of much of what has been described in this book. We began with the oxygen atom, noting that only the p states were in the energy range to make them important, but included the splitting of the energies into majority-spin and minority-spin states as indicated to the far

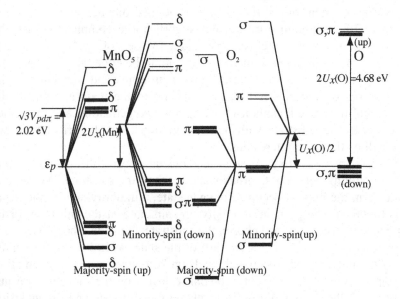

Fig. 8.6. Energy levels, all measured from the oxygen majority-spin p-state energy, for the MnO_5 surface cluster in $SrMnO_3$ on the left, for an oxygen molecule in the middle, and for the oxygen atom on the right, showing the majority-spin and minority-spin levels for each. Each line represents one level; closely-spaced lines are degenerate levels. The highest two and lowest two cluster levels for each case are e_g levels; the other cluster levels are t_g levels. All are distinguished as either σ, π, or δ. Those occupied in the ground state for the neutral oxygen atom, molecule, and for the Mn^{4+} clusters in $SrMnO_3$ are drawn heavy. In $LaMnO_3$ the upper majority-spin σ state is also occupied.

right in Fig. 8.6 and as discussed in Appendix 3C. Occupied levels are shown as heavy lines. We arbitrarily choose the majority spin in the oxygen to be the down-spin, the minority up.

We also obtain the electronic structure of the oxygen molecule as in Section 3.3, bonding and antibonding π states and σ states. We again include the exchange interaction which shifts the majority-spin states (again down-spin) down with respect to the minority-spin states, as for the next set of levels just to the left of the atomic levels in Fig. 8.6, relative to those next further to the left. Again, the occupied levels are shown as heavy lines; note in particular the majority-spin antibonding π states for O_2 which are both occupied with majority spin, one x-oriented and one y-oriented. The corresponding antibonding π states with minority spin are both empty. They provide the spin of two Bohr magnetons to the molecule.

We next constructed the cluster states for an $SrMnO_3$ substrate cluster just as in Fig. 8.2, but in this case a *surface* cluster. The Mn ion has four neighboring oxygen ions in the surface plane, and one below it, but no ion above it. This is straight-forward, but slightly more complicated than the fully symmetric octahedral cluster, giving the two sets of levels to the far left in Fig. 8.6. We could choose the majority spin to be either spin, but we find the strongest interactions between the oxygen atoms and molecules if the majority spin on the cluster is taken opposite to the majority spin of the oxygen, and we make that choice for Fig. 8.6. Oxygen atoms and molecules will come to the substrate with both orientations, but the orientations we chose will be the more interesting.

We next bring an oxygen atom or molecule toward the surface so that there is coupling between its orbitals and those of the substrate. We see that for the atom the highest occupied oxygen state, a minority-spin p state, spin up, is higher in energy than an empty up-spin state of the substrate. (That would also be true if we had chosen the substrate spins the other way.) It is essential to add the $U^* = U - e^2/d$ shift of the states as in Section 4.5. In the case of the molecule, the occupied levels on both sides are lower than all of the empty levels, but the empty up-spin π levels on the molecule are just barely above the occupied up-spin levels on the substrate and the inclusion of the U^* is again important.

We treated adsorption of both oxygen atoms and molecules, but here we restrict ourselves to a free oxygen atom and let it come down directly over a surface Mn ion. We can solve for the energy levels arising from the coupling between substrate and atomic levels, including the shifts for nonorthogonality, just as we have done in many other cases here. The results are shown in Fig. 8.7, analogous to the corresponding plot for the oxygen molecule as a function of oxygen spacing in Fig. 3.2. The interaction was largest if we occupied the up-spin σ state on the atom (rather than the π state of the same energy) since as the atom approached the substrate that energy dropped by more than two eV in energy as the atom approached, largely due to the decreasing U^*. The total energy dropped to a minimum total energy of -2.89 eV with the oxygen atom 1.8 Å above the substrate, as seen in the figure. There was no formal change in charge state, so we would call this a polar covalent bond of a neutral oxygen atom to the surface, with the oxygen actually losing some electronic charge to the Mn cluster. As would be expected, we did not find this strong bonding when the oxygen atom came down over a surface oxygen ion.

If this were $LaMnO_3$, rather than $SrMnO_3$, the one extra electron would occupy the lowest empty state, which is seen in Fig. 8.6 to be the majority-spin σ state in the cluster, the same state that dominated the binding of the O atom. In this case, we switched the relative spins of atom and substrate, and switched the minority-spin electron to a π state, so that a similar bonding can be achieved with the π state. The π coupling is weaker and this arrangement led to a weaker bonding of -1.51 eV at a distance of 1.86 Å. [In hindsight it is not clear whether we would have done better to keep the spins as for SRO, since the upper occupied majority-spin σ electron could be transferred to the nonbonding x^2-y^2 majority-spin state. We have not redone the calculation, but the central conclusions would seem to be the same.]

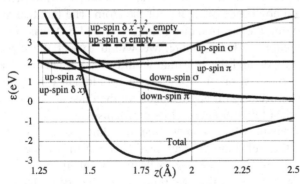

Fig. 8.7. Highest occupied states of each category for an O atom, as a function of the distance z to an Mn ion under it in an $SrMnO_3$ substrate. The total includes additional repulsions, is a minimum near 1.8 Å, and is measured from the energy at large distances. The cusps at 1.90 Å arise from $U^*(Mn) = 0$ for distances less than that, and are not physical. (After Harrison, 2010)

Our description of this oxygen atom on LSM contrasts strongly with that obtained from local density functional theory for the same system by Kotomin, Mastrikov, Heifets, and Maier (2008) and from LDA plus U by Lee, Kleis, Rossmeisl, and Morgan (2009). Both of these approaches give much more accurate representations of the bulk-crystal binding energy and equilibrium spacing than our approach. However, they both describe the system in terms of bands while it is clear from the small-polaron electrical conduction and magnetism that a local description is more appropriate. They both find the minimum energy with the oxygen over the Mn ion, as we did, but they find a considerable electronic charge transferred to the added oxygen from the substrate bands, with Lee, et al., describing it as an O^{2-} ion. Kotomin, et al., find an oxygen atom binding energy of -4.02 eV (relative to an isolated atom) at 1.63 Å, while Lee, et al., find a value of -3.2 eV (for their optimal U_{eff}) at an unspecified distance (compared to our of -2.89 eV at $z = 1.8$ Å). If we also do a *band calculation*, based upon our parameters, we also find broad, partly occupied bands and these much larger binding energies and charge transfers. It suggests that the difference is in our basic formulation, rather than our simplifying approximations. We proceed with our own representation of the electronic structure as we now discuss atoms arriving and leaving.

We return to an oxygen atom coming in over an Mn^{4+} site in $SrMnO_3$, described by Fig. 8.7. With thermal kinetic energy, near zero, it will be accelerated by the dropping total energy, acquiring a kinetic energy near 3 eV before being turned around near $z = 1.5$ Å and accelerated back outward, leaving the surface. With just what we have included so far, there can be no energy loss and no chance of capture of the atom. The fact that there was no barrier to reaching the position of a bound atom at 1.8 Å was not a sufficient condition for capture. Furthermore, we would expect the charge transfer between the oxygen and the substrate would be dominated by the occupied minority-spin σ state on the oxygen; it would shift partially onto the substrate cluster orbital as it approached, but then just as easily transfer back as it left. We must look for energy-loss mechanisms.

The mechanism which comes first to mind is a loss of energy to vibrations in the substrate. It can readily be estimated classically, and quantum effects are not expected to be important, using the total energy curve from Fig. 8.7. To do this, we represented the crystal by a chain of ten atoms, alternately Mn and O, as shown to the right in Fig. 8.8. They are connected by springs, with constants $\kappa = 16$ eV/Å^2 fit to the bulk modulus of SMO, within the chain and to four lateral neighbors with the same κ. The

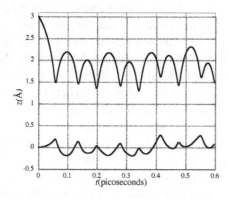 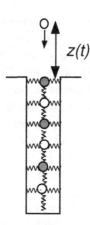

Fig. 8.8. The upper curve is the position of an O atom, initially approaching an Mn ion in a surface from directly above with 100 meV kinetic energy. The substrate was modeled as illustrated to the right, but with a chain of ten atoms, alternately Mn and O. Below on the left is shown the position of the top (Mn) ion of the chain as a function of time.

result of such a classical dynamical calculation is shown to the left in Fig. 8.8. Results were similar if we replaced the chain by a single atom, but then as energy is transferred into and out of the single mode, the oxygen atom is soon kicked off; with ten atoms this takes much longer as the energy was distributed in many modes. At least for these arrival parameters we expect a sticking coefficient near one for a single oxygen atom, leaving it bound over the Mn site at some −2.89 eV. The loss should be similar if it arrived at an oxygen site, but it would end up again over the Mn. In Harrison (2010) we found a similar loss pattern for an arriving oxygen molecule, leaving it bound at some −1 eV to the surface over an Mn^{4+} surface cluster, with lowest energy with its axis parallel to the surface.

The large loss to vibrations arose only because of the considerable acceleration of the oxygen atom, to 3 eV, before striking the surface like a hammer. With only the repulsive term in our fit to the energy as a function of z, almost no vibration is excited as the oxygen atom bounces off the surface.

We discussed a second very interesting mechanism for energy loss, which arises if among levels shifting with z as an atom arrives in Fig. 8.7, an occupied level crosses an empty level. The electron can be shifted to the empty state, and if it does not shift back as the atom leaves, that electronic energy will be left behind. This did not arise for the oxygen atom at accessible energies, but can for the molecule, and it can for LSM where we found such a crossing for an oxygen molecule coming in with its axis parallel to the surface. We noted in Harrison (2010) that two levels will cross only with this ideal arrival directly over the Mn ion. If slightly displaced laterally, the levels are split such that an arriving electron will transfer to the second level (in Fig 8.7, a dashed line, and actually the upper dashed line, representing a level uncoupled to the oxygen orbitals for ideal geometry), and transfer back when recoiling. However, there is a "sweet spot" near the ideal position (we estimate 5% of the surface area) within which it will most likely remain on the rising level. Thus it can be captured by entering the sweet spot at an angle and recoiling outside that spot, or entering outside and recoiling through the spot. Either way the electron remains in the higher-energy substrate level, capturing the atom or molecule, with the excess energy dissipated later.

In this case, a real charge transfer has occurred, with both the substrate and the atom or molecule changing their formal charge state. For Fig. 8.7, if the atom had sufficient energy to reach the crossing at 3.5 eV, a charged O^+ would have formed, which could later acquire an electron from the substrate to become the neutral oxygen bound to the surface at -2.89 eV. We went on in Harrison (2010) to see how these neutral species can roll into a neutral surface oxygen vacancy, incorporating the oxygen in the LSM which is serving as a cathode for a fuel cell, but we forgo that discussion here. Otherwise, the discussion is up-to-date on our continuing effort to understand the behavior of oxide fuel-cell cathodes.

earth, ground

Appendixes

1A The van-der-Waals Interaction and Atomic Polarizability

We seek the interaction between two atoms, or molecules, well separated from each other. Even though the electrons cannot move between the two atoms, so no bonding occurs, there is interaction between the electrons in the two atoms. We may write the distance between the two electrons as a vector $\mathbf{r} + \mathbf{r}_2 - \mathbf{r}_1$ as in Fig. 1A.1, where $\mathbf{r}$ is the position of the second nucleus relative to the first and $\mathbf{r}_1$ and $\mathbf{r}_2$ are the positions of the electrons measured from the nucleus of their atom. We chose a z axis along $\mathbf{r}$ and we may write out the separation in components, and the distances as the square root of the sum of the squared components. We then expand the resulting electron-electron interaction for small distances from the nucleus compared to the distance r between nuclei.

The leading term is just e^2/r which combines with the corresponding interaction between each electron and the other nucleus and between nuclei to give no Coulomb repulsion nor attraction between neutral atoms. Similarly, the terms linear in $\mathbf{r}_1$ or $\mathbf{r}_2$ are combined to show that the electron on one atom does not notice a neutral neighbor to this order. However, there

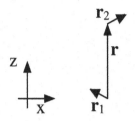

Fig. 1A.1. The coordinate system for writing the interaction between electrons on two atoms separated by **r**.

are terms bilinear in $\mathbf{r}_1$ and $\mathbf{r}_2$ which are of interest. They are a little intricate to calculate, but are straightforwardly obtained as a

$$V(\mathbf{r}_1,\mathbf{r}_2) = -\frac{3z_1z_2e^2}{r^3} + \frac{x_1x_2e^2}{r^3} + \frac{y_1y_2e^2}{r^3}. \tag{1A.1}$$

Now if we consider the *two-electron* wavefunction for these two electrons $\psi(\mathbf{r}_1)\psi(\mathbf{r}_2)$, for example both in the s ground state of their atom, there is a coupling to another two-electron state with both electrons in an excited p state of their atom, This two-electron coupling matrix element is the product of two one-electron matrix elements which for z_j will only be nonzero for the z-oriented p state and for that state we write the matrix element between the two states on one atom as

$$P_{sp} \equiv <p_z|z|s> \equiv \int \psi_{pz}(\mathbf{r})z\psi_s(\mathbf{r})d^3r. \tag{1A.2}$$

For the model wavefunctions of Eq. (1.2) (the p state has an extra factor of $z\sqrt{3}/r$) this is given by

$$P_{sp} \approx \frac{4\sqrt{3}(\mu_s\mu_p)^{3/2}}{(\mu_s+\mu_p)^4}. \tag{1A.3}$$

Then if one s state is occupied on each atom in the ground state, and the p states all empty, there is a lowering in energy given by the squared two-electron matrix element (the cross terms vanish), divided by the energy difference when both electrons are excited ($2(\varepsilon_p-\varepsilon_s)$), given by

$$E_{vdW} = -\frac{11P_{sp}^4 e^4}{4(\varepsilon_p - \varepsilon_s)r^6}, \tag{1A.4}$$

often called a van-der-Waals interaction after J. D. van der Waals who proposed interatomic interactions long before the advent of quantum mechanics. It is the principal interaction between neutral atoms at large distances.

This is correctly interpreted as arising from zero-point fluctuations of the electron cloud, inducing dipoles in the other atom, though this is not apparent in the derivation. For that reason it can be related to the *polarizability* of an atom, calculated in a similar way. An electric field **E** produces a perturbation $-e\mathbf{E}\cdot\mathbf{r}$ which lowers the energy of an atom, calculated just as described above, and given by $-e^2 E^2 P_{sp}^2/(\varepsilon_p - \varepsilon_s)$. This can be equated to the energy lowering of an atom of polarizability α by an electric field, $-\frac{1}{2}\alpha E^2$, from which the polarizability becomes

$$\alpha = \frac{2e^2 P_{sp}^2}{\varepsilon_p - \varepsilon_s}. \tag{1A.5}$$

Then the van-der-Waals interaction can be written in terms of the polarizability as

$$E_{vdW} = -\frac{11\alpha^2(\varepsilon_p - \varepsilon_s)}{16r^6}. \tag{1A.6}$$

In this form it can be applied to give the interaction between molecules as well as atoms, with suitable replacement of $\varepsilon_p - \varepsilon_s$ and other numerical factors.

Eq. (1A.5) provides a crude estimate of atomic polarizability. It is interesting to apply it to neon and other inert-gas atoms. There the valence p states are full, but there is an empty excited s state, so the same formula applies with the s and p states interchanged, and it is doubled since electrons of both spins contribute. We do not have the energy of the *excited s* state from Table 1.1, but we might guess it at half the p-state energy. Then $\mu_s = \mu_p/\sqrt{2}$ and Eq. (1A.3) yields $P_{sp}{}^2 = 0.235/\mu_p{}^2 = 0.235\hbar^2/(2m|\varepsilon_p|)$ and twice Eq. (1A.5) gives for the inert gases

$$\alpha = 0.94\frac{\hbar^2 e^2}{m\varepsilon_p^2}. \tag{1A.7}$$

We may even use it for helium assuming an excited p state half the ε_s and replace ε_p in Eq. (1A.7) by ε_s. These were compared with experiment in Table 1.2. We also used Eq. (1A.6) to estimate the cohesion of the inert gases and compared with experiment in the same table.

2A Calculation of Proton Positions

For helium, the $+2e$ nucleus and the $-2e\rho_0(r)$, from Eq. (1.3), form a neutral unit, so if at the distance r from the center there were no electron charge outside the radius r there would be no potential. The result will be useful elsewhere so we calculate the potential for a single electron of probability distribution $\rho_0(r')$ and its compensating central $+e$ proton. The potential which arises from a spherical shell of charge $dQ(r') = -e\rho_0(r')4\pi r'^2\,dr'$ for $r' > r$ is the potential it produces at all points inside that shell, $dQ(r')/r'$, minus the $dQ(r')/r$ at r it would produce if it were a spherical shell with radius less than r. Thus the potential due to the cloud of one electron charge density $-e\rho_0(r)$ without compensating nuclear charge is

$$\phi_0(r) = -e\int_r^\infty \rho_0(r')4\pi r'^2(\frac{1}{r'}-\frac{1}{r})dr' - \frac{e}{r} = e(\mu+\frac{1}{r})e^{-2\mu r} - \frac{e}{r} \tag{2A.1}$$

where the integrals could be done analytically and the result is generally negative, approaching $-e\mu$ at small r and $-e/r$ at large r, for this potential from a single electron. Thus the potential energy of the hydrogen molecule becomes the proton-electron interaction, $4e\phi_0(r)$ (for two protons r from the center, each seeing two electrons) plus the proton-proton interaction $e^2/2r$ plus the electron-electron interaction,

$$\int_{0,\infty} \rho_0(r)\phi_0(r)4\pi r^2 dr = \,^{11}/_8\, e^2\mu,$$ (2A.2)

the potential arising from one electron times the charge density of the other, integrated over all space. This last estimate may be much too large because the two electrons in fact correlate their motion to avoid each other. We should also add the kinetic energy of $\hbar^2\mu^2/2m$ for each of the two electrons with this wavefunction corresponding to Eq. (2.1). This leads to a prediction of the total energy of

$$E(r) = 4\left(e^2(\mu+\frac{1}{r})e^{-2\mu r} - \frac{e^2}{r} \right) + \frac{e^2}{2r} + \frac{11}{8}e^2\mu + \frac{\hbar^2\mu^2}{m}.$$ (2A.4)*

We use this as Eq. (2.2) for H_2 in the main text.

For the central hydrides we initially replace the helium $\mu = 2.56$ Å^{-1} by the neon $\mu = 2.46$ Å^{-1}, and with six p electrons rather than two s electrons the electron density becomes $6\rho_0(r)$ in terms of the $\rho_0(r)$ of Eq. (2.1). For hydrogen fluoride the loose proton then sees six times the potential from Eq. (2A.1), and has an energy $5e^2/r$ from interaction with the fluorine core, the only terms which depend upon r. In addition, the fluorine core of charge $+5e$ sees the potential of $-6e\mu$ from the six p electrons, and the electron-electron interaction becomes $\frac{1}{2} 6\times5\int_{0,\infty} \rho_0(r)\phi_0(r)4\pi r^2 dr$ from Eq. (2A.2) and the electron kinetic energy is from six electrons rather than two, leading to

$$E(r) = 6\left(e^2(\mu+\frac{1}{r})e^{-2\mu r} - \frac{e^2}{r} \right) + 5\frac{e^2}{r} - 30e^2\mu + 15\frac{11}{8}e^2\mu + 3\frac{\hbar^2\mu^2}{m}$$ (2A.5)

for HF. For the neon $\mu = 2.46$ Å^{-1} it is a minimum of -206 eV at $r = 0.93$ Å, and for the $\mu = 2.28$ Å^{-1} obtained from the fluorine p-state energy, it is a minimum of -200 eV at $r = 1.00$ Å, compared to the observed 1.08 Å, as discussed in the text.

We use this minimum energy of -206 eV in an unsuccessful attempt to obtain the cohesive energy. For isolated atoms the energy becomes $-25e^2\mu + 10\,^{11}/_8\, e^2\mu + 5\hbar^2\mu^2/2m$, evaluated with the fluorine μ, plus the -13.61 eV for H, for a total of -284 eV, even larger than the HF energy. We give up attempting the cohesive energy this way for the other central hydrides and

---

Note: Inadvertently, there is no Eq. (2A.3).

drop the final three constant terms in Eq. (2A.5), keeping only the two first terms in the counterpart of Eq. (2A.5). We postpone allowing the fluorine nucleus to move from the center of the charge distribution until we have examined the needed terms for water and ammonia.

Then for H_2O, the first term in Eq. (2A.5) is multiplied by two and from the final term to the right in Eq. (2A.1) there is a $-2 \times 6e^2/r$. There is also a $2 \times 4e^2/r$ for the two protons interacting with the $+4e$ oxygen core and an $e^2/2r$ for the two protons interacting with each other if they lie on the same diameter (linear H_2O), for

$$E(r) = 12\left(e^2(\mu + \frac{1}{r})e^{-2\mu r} - \frac{e^2}{r} \right) + 2 \times 4\frac{e^2}{r} + \frac{e^2}{2r} \qquad (2A.6)$$

for linear H_2O, plus constant terms.

If we allow the oxygen nucleus and the two protons to move freely in this spherical electron charge cloud, the interactions between them (the $8e^2/r$ and $e^2/2r$) change in a way that is easy to write down. In terms of the geometry illustrated in Fig. 2A.1, interaction between the protons and the nucleus becomes $8e^2/((b + r\sin\theta)^2 + (r\cos\theta)^2)^{1/2}$ and the $e^2/2r$ becomes $e^2/(2r\cos\theta)$ and the sum replaces the final two terms, $17e^2/2r$ in Eq. (2A.6).

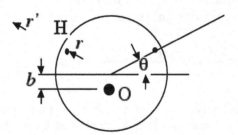

Fig. 2A.1. The geometry for the bent water molecule.

Writing the energy arising from the interaction of the oxygen nucleus with the electron charge cloud as it moves off center is more problematic. We cannot use same potential that we used for the protons because the variation of that potential with position at the center is directly related to the local charge density, which is quite inappropriate near the center. Further, freezing the charge density, with its cusp at the original nuclear position, would not seem to make sense as the nucleus moves. We therefore calculated the energy change as the nucleus moves away from the center based upon the electronic polarizability, which we calculated for Ne, in Eq. (1A.7) as $\alpha = 0.94e^2\hbar^2/m\varepsilon_p^2$. That polarizability could be interpreted in terms of an energy $\frac{1}{2}\kappa r^2$ as the neon nucleus moves with respect to the center of the charge cloud by $\mathbf{r}$, giving us an estimate of the κ for neon. For the polarizability arising predominantly from the six p electrons in neon, we would say a field $\mathbf{E}$ gives a force $\pm 6e\mathbf{E}$ and if the displacement energy is described by a spring constant κ, the relative displacement of that $6e$ charge will be $6e\mathbf{E}/\kappa$ and the dipole produced will be $36e^2\mathbf{E}/\kappa$ which we equate to $\alpha\mathbf{E}$. Thus the spring constant is given by $\kappa = 36e^2/\alpha = (36/0.94)m\varepsilon_p^2/\hbar^2$. For water we are assuming the same electron cloud but we should use the ε_p for oxygen, which best describes the electron cloud in water. Thus we could add the energy arising from this force,

$$E_{\text{nucleus}} = (18/0.94)m\varepsilon_p^2 b^2/\hbar^2 = 4.8\hbar^2\mu^4 b^2/m. \tag{2A.7}$$

With this weak restoring force we indeed find that the linear molecule is unstable and spontaneously deforms, but we find a displacement b of the oxygen of 0.14 Å beyond the range where the quadratic form, Eq. (2A.7), is meaningful. By $b = 0.14$ Å the restoring force based upon the model $\rho_0(r)$, which would have made the linear molecule stable, might be more appropriate. It seems that the soft restoring force, Eq. (2A.7), is correct and the linear molecule is unstable, but the force quickly becomes stronger, limiting the displacement. We should take this stiffening into account by adding a term in the energy proportional to b^4, a term which we might write as a constant times $\hbar^2\mu^6 b^4/m$, but we see no way to estimate that constant, other than fitting it to obtain the observed geometry. We adjust that coefficient and μ which is the only other parameter in the $E(r)$ to obtain the observed bend of $105°$ and the observed hydrogen-oxygen spacing of 0.96 Å. The resulting energy for distorted water becomes

$$E(r) = 2 \times 6\left(e^2(\mu + \frac{1}{r})e^{-2\mu r} - \frac{e^2}{r} \right)$$

$$+2\frac{4e^2}{\sqrt{(r\sin\theta + b)^2 + r^2\cos^2\theta}} + \frac{e^2}{2r\cos\theta} + \frac{9}{8}\frac{\hbar^2\mu^4 b^2}{m} + 203\frac{\hbar^2\mu^6 b^4}{m} \tag{2A.8}$$

with $\mu = 1.85$ Å^{-1} (compared to the oxygen value of 2.10 Å^{-1}). [With these adjustments, the values measured from the center of the charge cloud as in Fig. 2A.1 are $b = 0.065$ Å, $r = 0.92$ Å, and $\theta = 34.4°$ at the minimum energy, consistent with the observed O-H distance and H-O-H angle.] This μ is not so far from our predicted value and the adjusted b^4 term (1.1 eV) is similar in scale to the predicted b^2 term (−0.42 eV). Furthermore, our estimate of the b^2 term should be just as applicable to the other central hydrides, with the μ for the central atom in question, and we shall find that taking the same b^4 formula seems appropriate for ammonia, so we shall use it for all.

For ammonia, we may make the corresponding changes, but for three protons, again with a shift b of the nucleus. Using the counterpart of Fig. 2A.1, with θ again the angle between a line from the center of the distribution to a proton and the plane perpendicular to **b**, we obtain

$$E(r) = 3 \times 6 \left(e^2 (\mu + \frac{1}{r}) e^{-2\mu r} - \frac{e^2}{r} \right) + 3 \frac{3e^2}{\sqrt{(r\sin\theta + b)^2 + r^2 \cos^2\theta}}$$
$$+ \frac{3e^2}{\sqrt{3} r\cos\theta} + \frac{9}{8} \frac{\hbar^2 \mu^4 b^2}{m} + 203 \frac{\hbar^2 \mu^6 b^4}{m}. \tag{2A.9}$$

With the $\mu = 1.90$ Å^{-1} obtained for nitrogen, and all three protons in a plane with the nucleus, the minimum energy comes at a hydrogen-nitrogen distance of 0.81 Å, compared to the observed 1.01 Å. Using the same two final terms in b with the nitrogen μ the planar distribution is again unstable and the minimum energy is obtained with essentially the same hydrogen-nitrogen distance of 0.81 Å but $b = 0.06$ Å and an angle between proton-nitrogen distances of 110.5°, compared to the observed 107°. We may adjust the μ to 1.527 Å^{-1} to obtain the observed spacing at minimum energy with $b = 0.08$ Å and an angle between two oxygen-proton axes of 109°. One could improve this by tuning also the terms in b^2 and b^4, but we did not. It is gratifying that the values fit for water work so well here.

This also motivates us to use the same terms in b^n to allow the fluorine nucleus to move off-center. With the μ determined from the fluorine p-state energy we find the fluorine moving 0.042 Å and $d = 0.91$ Å. Adjusting μ to 1.92 Å^{-1} brought the spacing d up to the observed 1.08 Å with $b = 0.054$ Å.

Methane is simpler than ammonia, with a tetrahedral arrangement of the four protons. We obtain

$$E(r) = 4 \times 6 \left(e^2 (\mu + \frac{1}{r}) e^{-2\mu r} - \frac{e^2}{r} \right) + 4 \frac{2e^2}{r} + 6 \frac{e^2}{\sqrt{8/3}r} . \qquad (2A.10)$$

We went on to add terms to the energy when a molecule was near a metallic surface so that fields from image charges were felt by the molecule. For water, with $+4e$ at the position of the oxygen nucleus and $+e$ at each proton position, these extra terms were given by

$$\delta E(z) = -\frac{8e^2}{z+b} - \frac{2e^2}{2z - 2r\sin(\theta)} - \frac{2e^2}{2\sqrt{(z - r\sin(\theta))^2 + (r\cos(\theta))^2}}$$

$$-\frac{16e^2}{\sqrt{(2z + b - r\sin(\theta))^2 + (r\cos(\theta))^2}} + \frac{48e^2}{2z + b} \qquad (2A.11)$$

$$+\frac{24e^2}{\sqrt{(2z - r\sin(\theta))^2 + (r\cos(\theta))^2}} - \frac{36e^2}{2z}$$

in terms of the z of Fig. 2.10 and the b and θ of Fig. 2A.1. The last three terms involved the spherical distribution of $-6e$ charge, which could be replaced by a point charge at its center.

2B Radial Extension of s, p, and d States

We consider classical orbits in a Coulomb field (the same as in a gravitational field). We imagine two orbits of the same energy, but different angular momentum and therefore different eccentricity, as illustrated in Fig. 2B1. The circular orbit might be a p state and the elongated orbit an s state (though in quantum mechanics, with zero angular momentum for an s state it would pass through the nucleus). When they are at the same radial distance, and therefore at the same potential, as well as total energy, they must have the same kinetic energy. However, the s state has most of its kinetic energy associated with radial motion, so it will orbit far outside the circular orbit.

Correspondingly the wavefunction of an s state will extend far beyond that of the p state of the same energy. In the case of neon, this was more than compensated for by the fact that the p state had only half the energy,

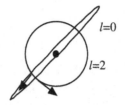

Fig. 2B.1. An orbit of high angular momentum ($\ell = 1$, or 2), with the same energy as one of low angular momentum ($\ell = 0$) will have the same speed where the orbits cross, but that with low angular momentum will be directed outward or inward, leading to a larger orbit.

corresponding to a circular classical orbit of twice the radius, larger than the average over time $<r>$ for the s state.

This effect is much more important for transition metals, with d states of two units ($2\hbar$) of angular momentum, but of energy quite close to that of the s states. Then the spacing between atoms will be set by the large s orbitals, and the much more localized d orbitals will have very weak interactions with their neighbors. This is even truer for f-shell metals with three units of angular momentum. In rare earths the orbitals are so close to the nucleus that they can be considered uncoupled from their neighbors.

The same relations of course apply to satellites of the earth. Their orbital acceleration is related to the angular velocity ω by $\partial^2 R / \partial t^2 = -\omega^2 R$. Near the earth's surface, where $R = 4000$ miles, the acceleration must be the gravitational acceleration, 32 feet/sec.^{2} so $\omega^2 = 32/(4000 \times 5280) = 1.52 \times 10^{-6}$ radian^2/sec^2, and the period $T = 2\pi/\omega = 5100$ sec. $= 1.4$ hours. That acceleration is related to the gravitational constant g by $\omega^2 R = gmM/R^2$ (with the satellite and earth masses of m and M), so the kinetic energy $\frac{1}{2}m\omega^2 R^2 = \frac{1}{2}gmM/R$ is half potential energy (also following from the Virial Theorem). If we kept the same energy, but eliminated the angular momentum by directing it outward, it would reach a radius of twice the earth's, 8000 miles, which could be Fig. 2B.1.

2C The Nature of the Hydrogen Bond

We may understand the structure of normal ice (called ice Ih) beginning with the zincblende structure of Fig. 5.1. Real ice would be based upon a structure which is locally the same as this, but with subtle differences in the symmetry of the stacking of subsequent planes of ions in the z direction, the same as the difference between fcc and hcp in Appendix 4B. [See Feibelman, 2010.] We ignore that difference. We imagine each of the circles in Fig. 5.1 as an oxygen ion (neon configuration of electrons) and place the two protons for each ion along the vector to its two neighbors below it in the picture. However, the electric dipole associated with every molecule is parallel, pointed downward. We could just as easily have it point upward by placing the protons above each oxygen, or pointing to the right or the left. In real water there is no net dipole. The proton between each pair of neighbors can tunnel between the two sites along the axis between them, just as an electron in a lithium molecule in Section 3.2 can tunnel from one atom to the other, in an electron state, shared between the two atoms. Here we describe the hydrogen bond as a proton state, shared between the two atoms. If we apply this view to the planar hexagonal pattern of Fig. 2.4, redrawn here on the left in Fig. 2C.1, each hydrogen-bond states can be represented by a dumbbell as to the right in Fig. 2C.1. Note that this uses 1½ protons per oxygen, allowing one hydrogen bond for each oxygen with a neighbor in the plane above or the plane below, exactly as for electron bonds in the zincblende structure (or wurzite structure (Feibelman, 2010)).

Though these hydrogen-bond states of the proton are qualitatively similar to electronic bond states, they are quantitatively very different. The tunneling of the proton in and out of a molecule in one bond state is strongly correlated with the tunneling of another proton in a bond state with that molecule, but that is less true for electron pairs in bonds. A similar difference in correlations can be seen in our treatment of the correlated electron bond in Section 3.5. If the coupling V_2 is made small, as arises for the larger mass of the proton, we found that the two electrons became segregated onto the two atoms, reducing the probability of two electrons being on the same atom. In ice, with small coupling, the probability of having *three* protons on one oxygen at the same time is similarly reduced.

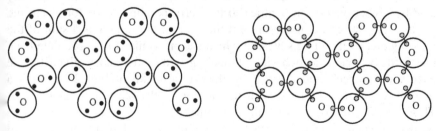

Fig. 2C.1, redrawn from Fig. 2.4. A schematic representation of the structure of ice, on the left with each proton near a neighboring molecule. As they tunnel between molecules they provide a hydrogen bond between the molecules, indicated by dumbbells to the right.

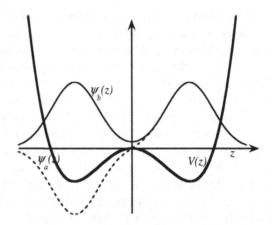

Fig. 2C.2. The heavy line is a schematic energy as a function of position for a proton forming a hydrogen bond. The light line is the ground (or bonding) state of the proton; the dashed line, an excited (or antibonding) state.

In many places in this book we seek to calculate the energy as a function of the position, z, of an atomic nucleus, in this case a proton. If we focus on a single hydrogen bond, that energy might look as the heavy line in Fig. 2C.2. We could solve the Schrödinger Equation for the wavefunction of a proton in this potential, which might look as the solid light line in Fig. 2C.2. We think of the state of the proton being just such a function when we discuss the hydrogen bond leading to the structure of ice; it is essentially a bonding state for the proton.

However, the antibonding state, obtained by changing the sign of the wavefunction to the left, shown as the dashed curve, gives a state of only slightly higher energy, in contrast to the corresponding states for electronic bonds. Thus we could also represent the state of the proton as a sum of the bonding and antibonding states, in which case it would slowly oscillate back and forth between the two sites, tunneling with a frequency given by the energy difference divided by $\hbar$. In the ice structure of Fig. 2C.1, such a motion would be correlated with the motion of the other hydrogen bonds. This second description would be much more appropriate for some properties, such as the outcome of experiments in which neutrons were fired at the ice, and we looked for the tracks of the protons knocked out of the bonds. For understanding the structure of ice, the first picture of equivalent sets of hydrogen bonds between every pair of neighbors, and no net electric dipole in that bond, is the appropriate one. Even for estimating the energy of the hydrogen bond, the classical description of a proton localized in one molecule, as would be done in density function theory, or as we did in our studies of the central hydrides in Section 2.2, would be a sensible approach. We did not undertake the calculation of the hydrogen-bond energy here.

3A Molecular Orbitals with Nonorthogonality

We give here more details of the formulation of molecular orbital theory for nonorthogonal orbitals, taken from Appendix B of Harrison (1989). We begin again with the expectation value of the Hamiltonian $<E> = <MO|H|MO>/<MO|MO>$ for two orbitals, which may differ in energy, given in Eq. (3.2). This is rewritten, defining $M_2 = <s_2|H|s_1> = <s_1|H|s_2>$, which we called V_2 before when the nonorthogonality $S = <s_2|s_1>$ was taken equal to zero, but we shall redefine V_2 here. We similarly define $M_3 = (<s_2|H|s_2> - <s_1|H|s_1>)/2$ in place of V_3 and an average s-state energy $M_4 = (<s_2|H|s_2> + <s_1|H|s_1>)/2$. Then Eq. (3.2) becomes

$$<E> = M_4 + \frac{(u_1^2 - u_2^2)M_3 + 2u_1u_2(M_2 - M_4)}{u_1^2 + 2u_1u_2S + u_2^2}. \tag{3A.1}$$

We note that M_4 is a constant and we take our zero of energy there, $M_4 = 0$. Then $<E>$ can be minimized with respect to u_1 and u_2. A convenient way to proceed is to minimize the numerator with respect to u_1 and u_2 independently, subject to the condition that the denominator is equal to one, applied with a Lagrange multiplier by adding $-\varepsilon(u_1{}^2 + 2u_1u_2S + u_2{}^2 - 1)$ to $<E>$.

We first do this for the simplest case where the nonorthogonality S is taken equal to zero, replacing the denominator by $u_1{}^2 + u_2{}^2$. Then the two minimization conditions yield

$$M_3u_1 + M_2u_2 = \varepsilon u_1$$

and $\tag{3A.2}$

$$M_2u_1 - M_3u_2 = \varepsilon u_2.$$

The condition that there be a solution is that

$$Det \begin{bmatrix} M_3 - \varepsilon & M_2 \\ M_2 & -M_3 - \varepsilon \end{bmatrix} = 0, \tag{3A.3}$$

which yields as a solution of the quadratic equation in ε,

$$\varepsilon = \pm \sqrt{(M_2^2 + M_3^2)}. \tag{3A.4}$$

$<E>$ takes one of those values, the maximum and minimum for $<E>$. These become our molecular energy eigenvalues, given by the average of the two atomic energies, minus or plus this square root of the coupling M_2 squared plus half the energy difference M_3 squared. This is the way we proceed in most of our analysis, but we next see the effect of a nonzero S.

We proceed just as above, but now keeping S and obtain in place of Eq. (3A.4),

$$\varepsilon = \frac{M_2 S \pm \left[M_2^2 S^2 + (1 - S^2)(M_2^2 + M_3^3) \right]^{1/2}}{1 - S^2}. \tag{3A.5}$$

Note that if S is taken to zero, it reduces to $\pm \sqrt{(M_2^2 + M_3^2)}$. Now we can redefine our covalent and polar energies as

$$V_2 = M_2/(1 - S^2) \tag{3A.6}$$

and

$$V_3 = M_3/(1 - S^2)^{1/2}. \tag{3A.7}$$

The energy eigenstates become

$$\varepsilon = \pm \sqrt{V_2^2 + V_3^2} + SV_2, \tag{3A.8}$$

measured from the average without coupling. The effect of the nonorthogonality is to modify slightly the meaning of the covalent and polar energies and to shift both the bonding and antibonding levels up by SV_2.

We may find a more suitable way to include this repulsion by turning to Extended Hückel Theory (Hoffmann, 1963), which is an alternative formulation of the coupling between neighboring orbitals. In that theory the matrix elements of the coupling were written as $V_2 = -KS(\varepsilon_1 + \varepsilon_2)/2$, with $S = \int \psi_1(\mathbf{r})\psi_2(\mathbf{r})d\tau$, the overlap between the two orbitals. The purpose was to find the coupling in terms of these overlaps which were readily computed.

As we pointed out at the beginning of Chapter 3, $K = 1.75$ was found to work well for the carbon-row elements. Thus it was another computational technique, simpler but less accurate than full local-density approximations. Our purpose is quite different, to use the proportionality between V_2 and S times the average orbital energy to write the shift

$$SV_2 = \lambda V_2^2 / \sqrt{(\varepsilon_1 \varepsilon_2)}, \qquad (3A.9)$$

since we independently have values for V_2. (For some reason the geometric mean of the two orbital energies seems more appropriate than an average.)

We may also look a little further into this alternative approach to coupling for comparison. We may in fact estimate the overlap S using the simple approximation to the wavefunction $\psi(\mathbf{r}) = \sqrt{(\mu^3/\pi)}\exp(-\mu r)$ of Eq. (1.1). This can readily be obtained numerically, giving the curve shown in Fig. 3A.1. Also shown is a simple exponential fit to this curve, which would be adequate over the range of interest here. We may also compare this with the free-electron formulae varying as $\hbar^2/md^2$. We find numerically that the calculated S of Fig. A3.1 is consistent with that free-electron formula $((d/S)\partial S/\partial d = -2)$ only at $\mu d = 3.33$ and a $1/d^2$ fit to the curve at that point gives the dashed curve also shown in Fig. 3A.1. These curves depend on which element is being considered only through the value of μ, and we can see that even if the free-electron fit is good at the large spacings in solids, it might greatly overestimate the couplings at the small spacings of diatomic molecules, as we found to be the case. Indeed the spacing of atoms tends to vary with the number X of nearest neighbors as $d \propto X^{1/4}$ (from Harrison (1999), p. 100 and as follows from Eq. (5.14)). Then with one

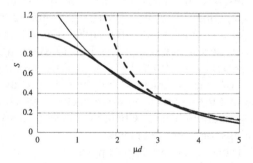

Fig. 3A.1. The overlap S of two exponential wavefunctions, each proportional to $\exp(-\mu r)$, but with centers separated by d. Also shown as the light line is an exponential fit to this curve given by $1.55\exp(-\mu d/2)$. The dashed curve is the "free-electron fit".

neighbor rather than the six we assumed from the free-electron fit we expect μd to be smaller for the molecule by a factor 0.64, taking it to $\mu d = 2.13$ in Fig. 3A.1. There the deviation of the free-electron fit is quite large.

Actually, for the oxygen molecule the μ for the p states is 2.09 Å^{-1}, so at the observed spacing $\mu d = 2.56$ and the "free-electron fit" is not far from the calculated S at that value for μd. On the other hand, the charge distribution for neon, shown in Fig. 1.2, obtained from the Hartree-Fock wavefunctions was smaller by a factor of two at large r than that calculated for p states from the exponential wavefunction of Eq. (1.1). This might suggest that an Extended-Hückel Theory based upon these Hartree-Fock wavefunctions would give smaller couplings by a factor of $1/\sqrt{2}$ than we obtained with our exponential form. In this context our scalings of the coupling in Table 3.1 do not look so unreasonable, except maybe for oxygen and fluorine.

3B Homopolar Molecular Orbitals

We defined the couplings in Fig. 3.1, choosing a z axis along the axis of the molecule, and p states which are proportional to z/r were designated by $|pz>$ as illustrated also in Fig. 1.1. They are called σ states because they have zero angular momentum around the internuclear axis, as do s states. There are also π states which are combinations of states with one unit of angular momentum around the internuclear axis. Those proportional to x/r are designated by $|px>$ and those proportional to y/r by $|py>$ with $|px>$ illustrated in Fig. 3.1. The figure also shows how the p states are divided into components to give values for the matrix elements when the orbitals are not aligned.

In solids we describe the states both as sums of atomic states, as here, and as freely propagating electrons with energies $\hbar^2 k^2/2m$. Matching the two descriptions for a simple-cubic lattice in Section 4.2 leads to couplings between the atomic states which we gave in Eqs. (3.8) through (3.11). In our treatments of solids (Harrison, 1989, 1999) we adjusted the coefficients to fit the energy bands of semiconductors, but here we retain the geometric coefficients. Note that the signs of the wavefunctions which are customarily chosen in each region are indicated. Changing those signs would change the signs of the couplings, or matrix elements. These couplings and the energies of the states from Table 1.1 are all that we need.

For homopolar molecules, the molecular orbitals can be chosen to be even (bonding) or odd (antibonding) under reflection in a plane bisecting their internuclear distance. Thus for an even state it will be a combination of the two states, $(|s_1\rangle + |s_2\rangle)/\sqrt{2}$ and $(|p_1\rangle - |p_2\rangle)/\sqrt{2}$, where the subscripts number the atoms, the denominator makes them normalized, and we have taken the signs for all atomic states as in the top row in Fig. 3.1. Then the calculation of the energies of the molecular orbitals would be a direct generalization of Eq. (3.5), except for the complication of the repulsive term from the nonorthogonality. We can also make a natural generalization of our treatment of that by defining energies for the two bond orbitals,

$$\varepsilon_{sb} = \varepsilon_s + V_{ss\sigma} + \lambda V_{ss\sigma}^2 / |\varepsilon_s|$$

and   (3B.1)

$$\varepsilon_{pb} = \varepsilon_p - V_{pp\sigma} + \lambda V_{pp\sigma}^2 / |\varepsilon_p|,$$

noting the signs of the couplings. Then the energy of the even molecular orbital is given by

$$\varepsilon_{MOb} = \frac{\varepsilon_{sb} + \varepsilon_{pb}}{2} \pm \sqrt{\left(\frac{\varepsilon_{pb} - \varepsilon_{sb}}{2}\right)^2 + V_{sp\sigma}^2 + \lambda V_{sp\sigma}^2 / \sqrt{\varepsilon_s \varepsilon_p}}, \qquad (3B.2)$$

with the minus sign giving the lowest-energy state of the molecule. There is also an upper bonding state, which might be written ε_{MOb}^*. Both of these are even under reflection in a plane bisecting the internuclear separation. [These are traditionally called *gerade* (even in German), but defined as reflection through a *point* at the center of the internuclear separation. Then for the π states which we discuss next, the bonding states are *ungerade* (odd). It seems an unnecessary complication to have "gerade" switch from bonding to antibonding.] There are also two antibonding states which are odd in reflection in plane, obtained by taking odd combinations for Eq. (3B.1),

$$\varepsilon_{sa} = \varepsilon_s - V_{ss\sigma} + \lambda V_{ss\sigma}^2 / |\varepsilon_s|$$

and   (3B.3)

$$\varepsilon_{pa} = \varepsilon_p + V_{pp\sigma} + \lambda V_{pp\sigma}^2 / |\varepsilon_p|.$$

These lead to upper and lower antibonding states with energies ε_{MOa} and ε_{MOa}^* obtained from the counterpart of Eq. (3B.2),

$$\varepsilon_{MOa} = \frac{\varepsilon_{sa} + \varepsilon_{pa}}{2} \pm \sqrt{\left(\frac{\varepsilon_{pa} - \varepsilon_{sa}}{2}\right)^2 + V_{sp\sigma}^2 + \lambda V_{sp\sigma}^2 / \sqrt{\varepsilon_s \varepsilon_p}} \,. \tag{3B.4}$$

Finally there are molecular orbitals based upon the p states oriented perpendicular to the internuclear axis, as to the right in the second row in Fig. 3.1. If the internuclear axis is taken as the z direction, there are bonding and antibonding states for p states with x orientation and molecular orbitals of the same energy for p states with y orientation. They are much simpler and are given by

$$\varepsilon_{\pi b} = \varepsilon_p + V_{pp\pi} + \lambda V_{pp\pi}^2 / |\varepsilon_p|$$

and   (3B.5)

$$\varepsilon_{\pi a} = \varepsilon_p - V_{pp\pi} + \lambda V_{pp\pi}^2 / |\varepsilon_p|,$$

noting that $V_{pp\pi} < 0$.

We proceeded with these for the other elements in the top row of the periodic table. For example, for Be$_2$ there are two electrons from each atom, which doubly occupy the two lowest states, the lower bonding state, Eq. (3B.2) with a minus sign, and the lower antibonding state, Eq. (3B.4) with the minus sign. From the sum of those four energies we subtract four times ε_s to obtain the cohesion.

It may be interesting to use these couplings with the state energies from Table 1.1 and to take $\lambda = 1$, before adjusting it as we did for Li$_2$, to predict the equilibrium spacing and the minimum total energies (an estimate of the cohesion), with results shown in Table 3B.1. Both the spacings and the cohesive energies are qualitatively correct, but neither very impressive. Lowering λ as we did for Li$_2$ in Section 3.2, to bring the spacing into accord, makes the predicted cohesion much too large. We attribute the discrepancies to our transfer of the free-electron couplings appropriate to the solid over to these closely-spaced molecules as discussed in Appendix 3A. Thus it seemed better to scale the couplings, as well as adjust the λ, to obtain both the spacing and the cohesion equal to the observed values, as we did in Section 3.3.

Table 3B.1. First estimates, $\lambda = 1$, and experimental values, of the spacing and cohesion for the diatomic molecules.

Molecule	Calc. d (Å)	Obs.d(Å)[a]	Calc.E (eV)	Obs.E(eV)[bv]
Li_2	2.83	2.67	−1.99	−1.1
Be_2	3.33	2.	−0.37	−0.7
B_2	3.09	1.59	−3.05	−3.
C_2	2.43	1.24	−6.55	−6.3
N_2	2.01	1.09	−11.92	−9.8
O_2	2.08	1.22	−8.78	−5.2
F_2	2.26	1.42	−5.55	−1.6
Cl_2	2.72	1.98	−3.85	−2.5
Br_2	2.86	2.28	−3.48	−2.44
I_2	3.05	2.66	−3.06	−2.22

[a]Slater (1968), Bader (1970), and Kittel (1966).
[b]Weast (1975), *Bond Strengths in Diatomic Molecules*, F215.

3C Exchange

Hund's Rule for atoms states that when two orbitals have the same energy, such as the x- and y-oriented p states in oxygen, the energy will be lowest if two electrons occupying them have the same spin. That will only be possible, because of the Pauli Exclusion Principle, if they each occupy a different orbital. In fact, the origin of the Pauli Principle and Hund's rule are the same, the required antisymmetry of the many-electron wavefunction: interchanging two electrons changes the sign of the composite wavefunction. That antisymmetry is in turn required (e.g., Harrison, 2000, p. 152) by quantum mechanics for particles of half-integral spin, such as electrons. Thus the wavefunction for two electrons (positions $\mathbf{r}_1$ and $\mathbf{r}_2$) of the same spin in two states (ψ_1 and ψ_2) can be written

$$\Psi(\mathbf{r}_1,\mathbf{r}_2) = \frac{\psi_1(\mathbf{r}_1)\psi_2(\mathbf{r}_2) - \psi_1(\mathbf{r}_2)\psi_2(\mathbf{r}_1)}{\sqrt{2}}. \tag{3C.1}$$

Note that it is equal to zero if $\mathbf{r}_1 = \mathbf{r}_2$. The electrons avoid each other and if the expectation value of the Coulomb repulsion, $e^2/|\mathbf{r}_1 - \mathbf{r}_2|$, is obtained with such an antisymmetric wavefunction it will be lower by some U_x than if the two electrons had antiparallel spin. For antiparallel spin, interchanging the spin changes the sign of the wavefunction so the orbital part of the wavefunction will be given by Eq. (3C.1) with the minus sign replaced by a plus. We *could* estimate U_x using atomic wavefunctions, but we take it from experiment.

For the oxygen atom, for example, the exchange energy is given by $U_x = 2.34$ eV (the energy lowering for each pair of p states with parallel spin on one oxygen atom, obtained from the NIST (National Institute of Science and Technology) tables of atomic spectra). That for oxygen is the main one we shall need. Values for other elements are given in Harrison (1999), and for d states in transition metals here in Table 7.2. This oxygen U_x shifts minority-spin levels in the atom by $2U_x$ since with minority spin the electron sees no other p states of the same spin, but with majority spin the electrons each see two p electrons with the same spin as its own. The molecular levels are shifted by only $U_x/2$ since the majority-spin electron spends only half its time on the same atom with the other majority-spin electron.

3D Polar Molecules

3D1 Cyanide, Full Calculation

We make a direct application of the theory we used for the homopolar molecules in Appendix 3B. to the polar cyanide molecule CN. The only choice comes in what to use for $\sqrt{(\varepsilon_s\varepsilon_p)}$ in Eq. (3.12). Since it is seen in Table 1.1 that the ratios of s- to p-state energies is about the same for all elements, we could have about the same value if we took the ε_s state from one atom and the ε_p state from the other, or the other way around. A use of $(\varepsilon_{s1}\varepsilon_{p2}\varepsilon_{s2}\varepsilon_{p1})^{1/4}$ is very reasonable.

We have values for the energies for all the atomic orbitals from Table 1.1, and all of the couplings as a function of d from Eqs. (3.8) through (3.11). For CN we average λ for C and N as 1.123 and the scale factor as 0.226 from Table 3.1. For any given d we shift the atomic orbital energies according to Eq. (3A.9) by λ times the squared coupling to each other orbital to which it is coupled, divided by $\sqrt{(\varepsilon_1\varepsilon_2)}$ for the two orbitals. Then doing the variational calculation for the expectation value of the energy of the orbital gives the counterpart of Eq. (3A.2) for a four-by-four Hamiltonian matrix, which can be written out as

$$f(\varepsilon) = (\varepsilon_{s1} - \varepsilon)(\varepsilon_{p1} - \varepsilon)(\varepsilon_{s2} - \varepsilon)(\varepsilon_{p2} - \varepsilon) - (\varepsilon_{s1} - \varepsilon)(\varepsilon_{p2} - \varepsilon)V_{sp\sigma}{}^2 - (\varepsilon_{p1} - \varepsilon)(\varepsilon_{s2} - \varepsilon)V_{sp\sigma}{}^2$$

$$\text{(3D.1)}$$

$$-(\varepsilon_{s1} - \varepsilon)(\varepsilon_{s2} - \varepsilon)V_{pp\sigma}{}^2 - (\varepsilon_{p1} - \varepsilon)(\varepsilon_{p2} - \varepsilon)V_{ss\sigma}{}^2 + V_{ss\sigma}{}^2V_{pp\sigma}{}^2 + V_{sp\sigma}{}^4 = 0.$$

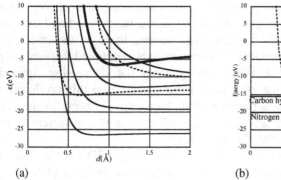
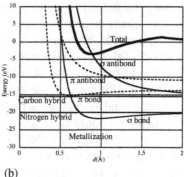

(a)   (b)

Fig. 3D.1. The energy levels of CN, in Part (a) based upon the full tight-binding calculation, the counterpart of Fig. 3.2, which showed the levels for O_2. The results in Part (b) are based upon hybrid states.

We wrote a small program which evaluated $f(\varepsilon)$ as ε varied for each value of d, noting when $f(\varepsilon)$ changed sign, to obtain the curves given in Part (a) of Fig. 3D.1, which includes also the π states dashed and the total energy as a heavy line.

In CN we have nine valence electrons per molecule (four for C and five for N), filling the two lowest σ levels with two electrons each, the π bonds with four electrons, and a single electron in the third σ state. The minimum in the total energy at $d = 1.10$ Å is somewhat removed from the observed spacing of 1.48 Å, but a moderate adjustment of λ from 1.22 to 0.97 and the scale factor from 0.249 to 0.52 leads to the observed spacing and bonding energy as indicated in Chapter 3.

3D2   CN with Hybrids

In Chapter 3 we also saw how to bypass the numerical procedure based upon Eq. (3D.1) by introducing hybrid states, and neglecting the coupling of the outward-leaning hybrid states with the states on the other atom. Then there were bonding and antibonding σ states based upon the two inward-leaning hybrids with a covalent and a polar energy. There were also nonbonding, dangling, hybrids. We obtain the hybrid state energies of $\varepsilon_{hN} = -20.03$ eV for N and $\varepsilon_{hC} = -15.23$ eV for C, using again Table 1.1, for a $V_3 = 2.40$ eV for the σ states, and we have $V_2 = 4.04\hbar^2/md^2$ from Eq. (3.15), equal to 14.05 eV at the equilibrium spacing of 1.48 Å. For the π states we have $V_3 = 1.385$ eV and $V_{pp\pi} = -(\pi^2/8)\hbar^2/md^2$. Using Eqs. (3.16) and (3.17) we obtain the levels shown in Fig. 3.3 and repeated here in Part (b) of Fig. 3D.1.

The dangling hybrids appear as horizontal lines independent of d. We again have nine electrons per atom pair. At small spacing the σ bond and the two dangling hybrids each accommodate two, leaving three in the π-bonds. Beyond $d = 1.7$ Å the antibonding σ becomes doubly occupied, leaving only a single electron in the π bond, arising from the decreasing upward nonorthogonality shift of the σ state at large spacing and this produced a cusp in the total energy as a function of d. This cusp is absent in the full calculation, but could have arisen with a small change in parameters since the σ and π states are calculated independently.

For the total energy based upon hybrid states in Fig. 3D.1 (b) we added the energies of the occupied states, and subtracted the energies of initially occupied atomic states to obtain a minimum energy of -3.66 eV at $d = 0.92$ Å, compared to the full calculation which gave -6.73 eV at $d = 1.10$ Å, closer to the observed -8.0 eV at 1.48 Å.

The principal difference was not the coupling between out-leaning hybrids and states on the *other* atom, which we neglected, but the *intra-atomic* coupling between the dangling hybrids and the antibonding states, which we discuss next.

3D3 Oxygen Metallization

We can most easily understand this by applying these simplifying formulae with hybrids to a homopolar case, O_2. Then for the σ states we have two dangling-hybrid levels at $\varepsilon_h = -25.4$ eV, a bonding and antibonding state $\varepsilon_h \pm V_2$ with again V_2 equal to the scale factor from Table 3.1 times $4.04\hbar^2/md^2$ (equal to 6.31 eV at the observed $d = 1.22$ Å for O_2), as well as the π states. If we doubly occupy the σ bond, the two dangling hybrids, four π bonds, and two π antibonds we obtain an energy per molecule, relative to free atoms, of

$$\varepsilon_{bond} = \varepsilon_p - \varepsilon_s - 2V_2 + 2\lambda V_2^2/|\varepsilon_h| - 2V_{pp\pi} + 6\lambda V_{pp\pi}^2/|\varepsilon_p|. \tag{3D.1}$$

The first two terms are a promotion energy in forming hybrids, as we find in semiconductors in Chapter 5, and if we move to very large spacing, and small coupling, we will have another level crossing and cusp as we saw in Fig. 3D.1 (b), so that the energy goes to zero at large spacing. Minimizing this energy with respect to spacing d at the small spacing, using the λ and scale factor for oxygen from Table 3.1, we find a spacing of 0.95 Å and $E =$

8.04 eV, a much larger discrepancy than for CN, eliminating the bonding altogether and leaving a positive result.

Comparing these simplified levels for O_2 with the full calculation in Fig. 3.2 we see that the two intermediate σ levels, corresponding to the dangling hybrids in the simplified approach, are split up and down by some ± 8 eV to become p-like and s-like levels. In the hybrid context this arises from a coupling between the inward-leaning and outward-leaning hybrid on the *same* atom, $<h|H|h'>$, which is negative and written as $-V_1$ with

$$V_1 = (\varepsilon_p - \varepsilon_s)/2 . \tag{3D.2}$$

This has the effect of pushing an even combination of hybrids up near the p-state energy and pushing the bond down by the same amount; that does not matter since both are occupied. However, it also pushes the occupied odd combination of hybrids down near the s-state energy and pushes the antibond up by the same amount; with the antibond empty it gives an energy gain equal to twice (for electrons of both spins) the downward shift $-V_1$ of the hybrids. The corresponding effect arises for bonds in semiconductors and we called V_1 the *metallic energy* and the effect the *metallization of the bond* (e.g., Harrison, 1989 and 1999), though here it might better be called a *dehybridization* energy. In solids it is a rather small contribution, which allows the bond to relax into the environment, as illustrated on the cover of Harrison (1989). In oxygen, with this metallic energy larger than the covalent energy, it brings into question the use of hybrids in the first place. We did better by calculating s bonds and antibonds and p bonds and antibonds, and then adding the effects of sp coupling afterward. However, we proceed further in order to learn about polar systems.

For the case of O_2 here we may evaluate the two states obtained from the coupling of the antibonding state (at $\varepsilon_h + V_2$ with the scaled $V_2 = 2.88$ eV at the observed spacing) to an odd combination of the outward-leaning hybrids (at ε_h) as

$$\varepsilon = V_2/2 \pm \sqrt{V_2^2/4 + V_1^2} \tag{3D.3}$$

relative to the levels without metallization. Note that with this coupling based upon the on-site energy difference there is no repulsive term with a λ. Doubly occupying the lower state of Eq. (3D.3) and leaving the upper one empty gains an energy

$$\varepsilon_{metal.} = V_2 - \sqrt{V_2^2 + 4V_1^2} \tag{3D.4}$$

relative to what it would be without this intra-atomic coupling V_1. Adding this to the energy of Eq. (3D.1) based upon the beginning λ and scale and minimizing with respect to d moves the spacing out to 1.11 Å and reduces the energy to -4.14 eV. This recovers most the error from dropping the coupling of the dangling hybrids; the λ and the scale could be adjusted to make up the rest.

When this arises in semiconductors, the corresponding V_1 is generally much smaller than the V_2 so the use of hybrids is a much better approximation. Then Eq. (3D.4) can be expanded for small V_1 to $\varepsilon_{metal.} = -2V_1^2/V_2$. That is not a good approximation here.

3D4 Metallization of CN

The calculation of metallization for CN is made much more complicated by the polarity. The antibonding σ state has a larger coefficient on the carbon hybrid, $\sqrt{((1+\alpha_p)/2)}$, than that on the nitrogen, $-\sqrt{((1-\alpha_p)/2)}$, with α_p given by Eq. (3.20). Similarly, there is one combination of the dangling hybrids which is uncoupled from the antibond and the other has coefficients equal to those from the antibond.

Again, for CN we have a total of nine valence electrons (4 from C and 5 from N) and if we look at the ordering of levels in Fig. 3D.1(b) we should fill the bonding σ state with 2, each dangling hybrid with two, leaving three for the bonding π state. This appeared in the end to be true, though if one looks at the full calculation in Fig. 3D.1 (a) we find the bonding π level slightly below the third σ level, corresponding to the carbon hybrid, suggesting we should leave one carbon dangling hybrid empty. We do not do that, and if both dangling hybrids are full, we include only their coupling with the empty antibonding σ state. We may then calculate the effect of the metallic energy V_{1N} on the nitrogen and find in analogy with Eq. (3D.4) a metallization energy of

$$\varepsilon_{metal.N} = \varepsilon_{MO}(\sigma,a) - \varepsilon_{hN} - \sqrt{(\varepsilon_{MO}(\sigma,a) - \varepsilon_{hN})^2 + 2V_{1N}^2(1-\alpha_p)} . \tag{3D.5}$$

Similarly we find for the V_{1C},

$$\varepsilon_{metal.C} = \varepsilon_{MO}(\sigma,a) - \varepsilon_{hC} - \sqrt{(\varepsilon_{MO}(\sigma,a) - \varepsilon_{hC})^2 + 2V_{1C}^2(1 - \alpha_p)} . \qquad (3D.6)$$

With this choice of occupation of the states, adding the metallization brought the predicted spacing to 1.53 Å and a bonding energy of −6.63 eV, close to the observed 1.48 Å and −8.0 eV. It also gave the carbon dangling-bond state at −18.06 eV compared to the carbon π bond at −14.02 eV consistent with that choice of occupation. Agreement was not nearly so close with the choice of one dangling hybrid empty. This seemed convincing support for that choice of occupation. Shifting to $\lambda = 1.055$ and a scale of 0.51 for the couplings would bring it into agreement, with the carbon dangling hybrid still convincingly below the bonding π states.

4A Lithium Energy Bands

We look a little more completely at the bands, focusing on the metal lithium. We again put it in a simple-cubic structure, with spacing 3.03 Å from the k_F of Table 4.1 and the couplings given in Eqs. (3.8)-(3.11). If we use the $\varepsilon_s - \varepsilon_p = \pi^2 \hbar^2 / md^2 = 8.19$ eV from the free-electron fit, proceeding as we shall describe, we obtain the bands of Fig. (4.4). We do not have a p-state energy for lithium in Table 1.1, but an extrapolation from other elements would suggest perhaps 1/3 of the ε_s leading to $\varepsilon_p - \varepsilon_s = 3.56$ eV. Here we should look to *ab initio* calculations or experiment for guidance. Perdew and Vosko (1977) have studied the lithium bands carefully and considered other calculations as well. They of course used the observed body-centered-cubic (bcc) structure and found a band gap of 2.8 eV at the nearest Zone face, near the slightly distorted Fermi sphere, as ours would be. This should be close to twice the back-scattering pseudopotential matrix element for electrons at the Fermi surface, as should our band gap at the Zone face, so we adjust to that gap, requiring $\varepsilon_s - \varepsilon_p = 5.39$ eV, midway between the two estimates, and we use that value here, taking ε_s from Table 1.1.

The calculation of the σ bands, as we did for Fig. 4.4, is straightforward. We had the s-band energies ε_k^s in Eq. (4.5) and we are evaluating them for $k_y = k_z = 0$, as a function of $k_x = k$. The p-band energies ε_k^p are of exactly the same form, but with the negative $V_{ss\sigma}$ replaced by the positive $V_{pp\sigma}$, three times as large in magnitude, and ε_s replaced by ε_p. The coupling between the two comes from a $V_{sp\sigma}$ coupling which can be seen to be of opposite sign for neighbors to the right and to the left, so that we obtain a coupling between the two band states of $\pm 2iV_{sp\sigma}\sin(kd)$ rather than a cosine. The resulting bands are obtained, in analogy with Eq. (3.3), as

$$\varepsilon_k = \frac{\varepsilon_k^s + \varepsilon_k^p}{2} \pm \sqrt{\left(\frac{\varepsilon_k^s - \varepsilon_k^p}{2}\right)^2 + (2V_{sp\sigma}\sin(kd))^2} \ . \tag{4A.1}$$

We have omitted a term $\lambda V_{sp\sigma}{}^2 / \sqrt{(\varepsilon_s \varepsilon_p)}$ as we omitted the corresponding terms in ε_k^s and ε_k^p, partly because of the uncertainty in dealing with ε_p. The bands have become quite free-electron-like and it would seem better to simply shift the entire band structure by the constant $3.37\hbar^2/md^2 = 2.80$ eV, which was the shift which cancelled half of the bonding energy gain leading to Eq. (4.6). The only reason for keeping such a shift will be in comparing energy levels with those of the water molecule, where such shifts are included.

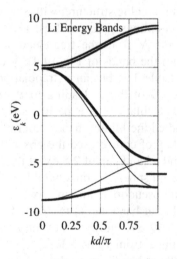

Fig. 4A.1. The energy bands for **k** in a [100] direction in simple-cubic lithium, calculated here. The thin lines are ε_k^s and ε_k^p and the full lines are the resulting σ bands. The upper bands are π bands, doubly degenerate. The surface band is shown as a short line to the right. All have been shifted up by $3.37\hbar^2/md^2$.

The energies of the π states are also of exactly the form of Eq. (4.5) but with the $V_{ss\sigma}$ replaced by $V_{pp\pi}$ which is equal, and ε_s replaced by ε_p. They are all readily evaluated with the results shown in Fig. 4A.1. [The σ bands here look much like the bands given by Perdew and Vosko for wavenumbers in the direction of the nearest Zone face for bcc lithium, but their π bands bend downward, rather than upward, due to interaction with other bands from above.]

We can see by the light lines that the p-like state lies lower at the Zone boundary, $kd = \pi$, so there will be a surface band mid-gap, as described in Section 4.5. It is designated by the short line to the right in Fig. 4A.1. Its energy varies in the lateral directions, with k_y and k_z. The lower band in Fig. 4A.1 is ordinarily thought of as an s band, but it is in fact p-like near this Zone boundary.

We saw that for a pure tight-binding s band for lithium the Fermi surface (the polyhedron in Fig. 4.3 intersected the Brillouin Zone, so that the states were occupied at the center of the Zone face. The free-electron surface is spherical and was completely inside the Zone, so we expect the Fermi surface for the bands in Fig. 4A.1 to be more nearly spherical and possibly not touch the Zone face. Perdew and Vosko (1977) indicated that the Fermi surface did not touch the Zone faces in the bcc lithium. In any case, we expect the Fermi energy to lie below the dangling-hybrid surface band indicated in the figure, leaving it empty. That question is central to the discussion of molecules at the surface in Section 4.5. In a simple-cubic divalent metal, the bands would be similar, but with a Fermi surface large enough to accommodate all of the electrons. These states in the second band at the Zone face are the lowest states in that band. They will certainly be occupied or the system would be insulating. The surface states below them will also be doubly occupied.

These bands are given on the scale of the atomic term values from Table 1.1, including shifts from nonorthogonality, as are the molecular orbitals of the water, as we indicated. However, the relative scales are not reliable. When two systems come into contact there is often charge transfer between them, from the formation of polar interface bonds, which shift the scales on the two sides relative to each other. That is the determining factor at junctions between semiconductors (e.g., Harrison, 1999, 741ff).

4B Metallic Crystal Structures

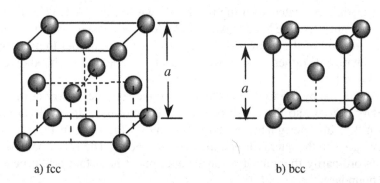

a) fcc   b) bcc

Fig. 4B.1. In part (a), a unit cube of the face-centered-cubic structure with eight cube-corner atoms and six face-center atoms. When the cube is repeated as a three-dimensional array, sharing the corner and face atoms, the two sites are seen to be equivalent, just depending upon where the cube is drawn. The density of atoms is $4/a^3$. In part (b), a unit cube of the body-centered-cubic structure, with atom density $2/a^3$.

We focused upon simple-cubic structures for lithium since they are so simple to picture and to treat, but they rarely occur in nature. The real crystal structures of metals are more closely packed, have a higher density of atoms. A very common structure, and the one which comes up most in this book, is the face-centered-cubic (fcc) structure. A unit cube of this fcc structure is shown in Fig. 4B.1(a). Every atom is equivalent so there are *primitive cells* each containing only one atom, into which the crystal could equally well be divided, but the larger cubic cell is convenient. Another common structure is body-centered-cubic, shown in Part (b), to which the same comment applies.

Another way of constructing the fcc structure is to start with a plane of closely-backed balls, having hexagonal symmetry, called an A plane. A second identical plane can be placed on the first, with each ball resting on three in the substrate, called a B plane. There was actually another way to do this with each ball on three atoms which form a triangle inverted with respect to that we chose, called a C plane. If we stack the planes as ABCABCA... we form the fcc structure, with the cube diagonal [111] direction normal to the planes, and this can be seen by studying Fig. 4B.1.

Stacking them ABABA... forms the third common structure, hexagonal-close-packed (hcp), clearly equally densely packed balls, but with a unique c *axis* perpendicular to the plane.

4C Water Molecules Bonding to Metals

We now treat the water molecule in terms of interatomic coupling between hydrogen and oxygen, as in Section 3.7. We let the oxygen-hydrogen vectors be at 90° as predicted there, and use the scaled couplings from Table 3.2. Then with $d = 0.96$ Å we obtain $V_{sp\sigma} = 3.90$ eV and $V_3 = (\varepsilon_s(\text{H})-\varepsilon_p(\text{O}))/2 = 1.58$ eV. Then the levels (four of which were added to get Eq. (3.27)) become,

$$\varepsilon = \frac{\varepsilon_s(H)+\varepsilon_p(O)}{2} \pm \sqrt{V_{sp\sigma}^2+V_3^2} + \frac{V_{sp\sigma}^2}{2\sqrt{V_{sp\sigma}^2+V_3^2}}. \qquad (4C.1)$$

These are shown in the energy-level diagram of Fig. 4B.1, along with the bands near the Brillouin Zone boundary from Appendix 4A. The highest

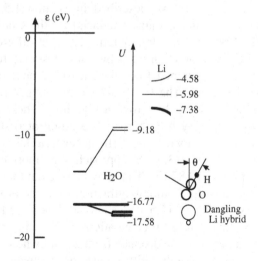

Fig. 4C.1. Important energy levels for water, and the bands near the Zone boundary for simple-cubic lithium. That at −5.98 eV is the dangling hybrid surface band. Occupied states are drawn heavy, empty states light. To the lower right is shown the coupling between the dangling hybrid and one H_2O state at −9.18 eV + U.

and lowest water levels are doubly degenerate and each would be split if we used the observed angle of 104.5° for water. Since they are degenerate we can choose any orthogonal combination and the simplest may be with one p state oriented toward one hydrogen as to the right in Fig. 4C.1, and the other p state oriented toward the other, not shown in the figure, but along the second light line perpendicular to the first. We should also note the polarity of the bonding within the molecule, $\alpha_p = V_3 / \sqrt{(V_{sp\sigma}^2 + V_3^2)} = 0.38$, so the coefficient on the oxygen p state for each *antibonding* (upper) state from Eq. (3.22) is $\sqrt{((1 - \alpha_p)/2)} = 0.56$ and $\sqrt{((1 + \alpha_p)/2)} = 0.83$ on the hydrogen s state to which it is coupled. We have also shown the level for the p state perpendicular to the plane of the molecule, nonbonding and occupied at the oxygen p-state energy -16.77 eV.

We cannot work with these quite as directly as we could the other levels we have described up to this point. We have found occupied levels in the metal, higher in energy than an empty levels in the water molecule. We encountered this difficulty also in the case of oxygen molecules approaching a metal-oxide fuel-cell cathode (Harrison (2010)), and we use the approach we took there.

The energy levels in Table 1.1, and the levels we have calculated here represent the energy to remove an electron from the atom or molecule, or to return that same electron, as we described in Section 1.5. If we add an *additional* electron, the energy gain is reduced by the Coulomb repulsion between it and the first electron already returned. That difference, which we called U, is of the order of e^2 divided by some distance representing the atomic size and values obtained from the measured spectra of the atoms were tabulated in Table 1.1. The $U = 14.47$ eV for oxygen raises the empty oxygen levels as indicated by the arrow in Fig. 4C.1. These are huge shifts, but they have come up very little before. We might at first think that this should have come up when we formed HF in Section 3.7, with a hydrogen atom with an electron at -13.6 eV approaches a fluorine atom with an empty p state with $\varepsilon_p = -19.87$ eV. That may be true but that extra energy $U = 15.0$ eV is reduced by the Coulomb attraction from the proton left behind, a distance d away, $-e^2/d = -15.6$ eV, so we did not need to take it into account. This was usually true for calculations of bonding in molecules and solids, but not for an atom some distance from a solid where the cancellation is reduced and the Coulomb shifts must be included. We take the cancellation within the molecule to be quantitative, so may take the U for the molecule to be e^2 divided by the spacing within the molecule, which for water gives 15 eV, essentially the same as the oxygen value from Table 1.1.

As a water molecule approaches a surface, there is again a term canceling against this U, but in the case of a metal this would seem to be the image potential from the metal if an electron was transferred to the molecule, and this lowering in energy is only $-e^2/(4z)$ if the molecule is a distance z from the surface, -3 eV at 1 Å. [The force due to the image is $e^2/(2z)^2$, and the energy is obtained integrating the force times dz from large distances.] We expect the energy at which an electron can be added to the molecule to remain far above the occupied states, and have little contribution to the energy from coupling with the substrate. In fact the coupling with the occupied states, based upon inward-leaning hybrids will be very weak, and the strong coupling to the outward-leaning dangling hybrid is coupling between states both empty, and not of consequence. The electrostatic effects which we described in Section 2.6 may be the dominant terms.

This may not be true if the metallic substrate is divalent or polyvalent. The bands will be similar to Figs. 4A.1 and 4C.1, but the Fermi energy will lie in the upper band shown to the right and the dangling hybrid will be doubly occupied. Then clearly the dominant interaction between the water molecule and the substrate will be between the empty antibonding state of the molecule and the occupied dangling hybrid on the substrate.

The central question, once we know that the substrate dangling hybrid and the antibonding state are the important levels, is whether the coupling is stronger with the oxygen or with the hydrogen leading. The antibonds are based upon an oxygen p state and a hydrogen s state, and the coupling with the substrate will be strongest if the bond axis is along the normal to the surface at the site of the dangling hybrid on the substrate. The coefficient of the oxygen for the *antibond* on the water is $-\sqrt{((1-\alpha_i)2)} = -0.56$ since the oxygen is the orbital of lower energy (Eq. (3.22)). The coefficient on the hydrogen is $\sqrt{((1+\alpha_i)2)} = 0.83$. We may also use our matrix elements from Eqs. (3.8) through (3.11) to find the coupling of a hybrid $(|s> + |p>)/\sqrt{2}$ to a hydrogen s state as $3.49\hbar^2/mz^2$ at a distance z, and to a p state as $3.73\hbar^2/mz^2$, so that the coupling through the oxygen p state has magnitude $2.09\hbar^2/mz^2$ and the coupling through the hydrogen s state is $2.89\hbar^2/mz^2$. An arrival with the hydrogen between the oxygen and the metal is favored by this bonding, as it was favored by electrostatic considerations, and as Feibelman (2010) indicates is correct.

5A Bonding in SiO₂

Pantelides and Harrison (1976) postulated a bonding unit in silicon dioxide with two silicon sp^3 hybrids with $\varepsilon_h = -9.39$ eV coupled to oxygen p states, as in Fig. 5A.1, initially neglecting coupling to bonding units involving other oxygen atoms. There are states ε_x even in reflection through the x axis and states ε_z odd in that reflection with energies given by

$$\varepsilon_x = (\varepsilon_h + \varepsilon_p)/2 \pm \sqrt{(\varepsilon_h - \varepsilon_p)^2/4 + 2V_{hp}^2 \sin^2 \theta}$$

and   (5A.1)

$$\varepsilon_z = (\varepsilon_h + \varepsilon_p)/2 \pm \sqrt{(\varepsilon_h - \varepsilon_p)^2/4 + 2V_{hp}^2 \cos^2 \theta},$$

before introducing nonorthogonality shifts. Here $V_{hp} = V_{sp\sigma}/2 + \sqrt{3}V_{pp\sigma}/2 = 3.99\hbar^2/md^2 = 11.72$ eV for our couplings and $(\varepsilon_h - \varepsilon_p(O))/2 = 3.69$ eV. The bonding levels for $\theta = 18°$ become $\varepsilon_x = -19.40$ eV and ε_z is -29.28 eV and the π state, oriented perpendicular to the figure, remains at $\varepsilon_p(O)$. The corresponding energy-level diagram is shown in Fig. 5A.2 (also as Fig. 5.9). We gain energy from two electrons in ε_x and two in ε_z and the total energy per bonding unit is minimum with $\theta = 45°$, so that $\mathbf{d}_1$ is perpendicular to $\mathbf{d}_2$. That explains why the bonding unit is bent, and other effects appear to reduce θ to $18°$ (discussed in Harrison (1999) 412ff). Using the observed $\theta = 18°$ leads to the energy levels given above. For the cohesion, we first promote the silicon electrons from $2(\varepsilon_s + \varepsilon_p)$ to $4\varepsilon_h$ costing 7.20 eV, or half

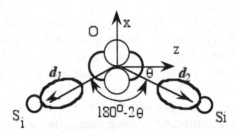

Fig. 5A.1. Notation for the bonding unit containing an oxygen atom and its bonds to two neighboring silicon atoms. $d_1 = d_2 = 1.61$ Å.

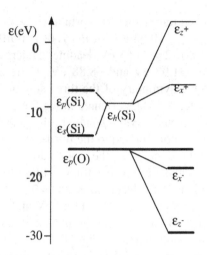

Fig. 5A.2. The energy-level diagram for SiO_2, from Fig. 5.9.

that per bonding unit. Then two electrons from that hybrid and two oxygen p electrons per bonding unit drop into bonding states, gaining -45.04 eV per bonding unit. There is also the repulsion which we take as C/d^4 in each of the two bonds and set the derivative with respect to d of the bonding energy plus this $2C/d^4$ equal to zero, and we obtain a total repulsion

$$2\frac{C}{d^4} = \frac{2V_{hp}^2 \sin^2 \theta}{\sqrt{(\varepsilon_h - \varepsilon_p)^2/4 + 2V_{hp}^2 \sin^2 \theta}} + \frac{2V_{hp}^2 \cos^2 \theta}{\sqrt{(\varepsilon_h - \varepsilon_p)^2/4 + 2V_{hp}^2 \cos^2 \theta}} = 19.50 \text{ eV}$$

(5A.2)

eV for a total cohesion of -21.94 eV per bonding unit, compared to the experimental -9.0 eV. This could not be brought into agreement by adjusting the coupling V_{hp} since even at zero, the transfer energy of $2\varepsilon_p(\text{O})$ $-\varepsilon_s(\text{Si}) - \varepsilon_p(\text{Si}) = -18.75$ eV exceeds the experimental value. The probable resolution comes in Chapter 6 where we discuss corrections to the transfer energy in oxides.

5B States in the Sulphate Ion

As indicated in the text, in the sulphate ion our bonding unit becomes the sulphur atom with a surrounding tetrahedron of oxygen ions, as to the right in Fig. 5B.1. Then there is one state (a cluster orbital) with the full symmetry of the tetrahedron, containing the sulphur s state, and radially oriented p states on each ion, (a cluster p orbital $(|p_1> + |p_2> + |p_3> + |p_4>)/2)$ described by a $V_2 = 2V_{sp\sigma} = 10.93$ eV for $d = 1.48$ Å (Gillespie, 1972, p. 150) and a $V_3 = (\varepsilon_s(S) - \varepsilon_p(O))/2 = -3.63$ eV, leading to energy levels -20.40 eV $\pm \sqrt{(V_2^2 + V_3^2)}$, equal to -31.92 eV and -8.88 eV. There are also three cluster orbitals which have the symmetry of the three sulphur p states. They are a little trickier to evaluate. A p state oriented along the z axis as in Fig. 5B.1 is coupled to a radially oriented oxygen p state by $V_{pp\sigma}\cos\theta$ and to a tangential p state by $V_{pp\pi}\sin\theta$ (and not coupled to the third). We make a cluster orbital, as for the cluster orbital coupled to the s state above (but with appropriate signs for the p states) and must solve for the two cluster orbitals and the central p state together. This leads to a $V_2^2 = 2V_{pp\sigma}^2\cos^2\theta + 2V_{pp\pi}^2\sin^2\theta$ (with $\cos^2\theta = 1/3$, $\sin^2\theta = 2/3$), or $V_2 = 11.62$ eV and $V_3 = 2.59$ eV, leading to levels $-14.19 \pm \sqrt{(V_2^2 + V_3^2)} = -26.09$ eV and -2.28 eV. The corresponding energy-level diagram is shown here in Fig. 5B.1 and in Fig. 5.10.

To obtain the cohesion for $MgSO_4$ we first subtract the energies of the atomic states initially occupied, two in Mg s, two in S s, and four in S p states from the energy of the eight occupying the four bonding states in Fig. 5B.1 to obtain -112.16 eV. For the repulsion we add three times the counterpart of Eq. (5A.2) for the states based upon sulphur p states to the corresponding counterpart based upon the sulphur s state to obtain 43.78 eV. The predicted cohesive energy per molecule then is -68.38 eV, compared to

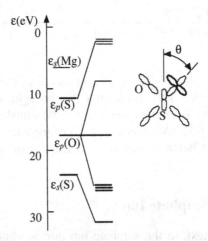

Fig. 5B.1. Energy levels for a sulphate ion from $MgSO_4$. One π-oriented p state from the oxygen tetrahedron is shown with heavy lines in the sketch to the right.

the experimental value of -28.01 eV. The experimental value was obtained from the enthalpy of formation of $MgSO_4$ of $\Delta Hf^\circ = -305.5$ kcal/mole $= -13.25$ eV per molecule, which we obtained from the CRC Handbook (Weast, 1975). That is the energy of formation of the compound from its elements at 25° C. We then subtracted the energy to dissociate the elements into atoms at $0^\circ K$ from Kittel (1976), 1.51 eV per atom for Mg, 2.85 eV per atom for sulphur and four times 2.6 eV for oxygen, neglecting the difference in temperature for the two contributions.

The calculation of the states for $NaClO_4$ is just the same, with different term values for the different elements and with a spacing of 1.42 Å. These lead to s-like states with a $V_2 = 11.87$ eV and a $V_3 = -6.22$ eV and p-like states with a $V_2 = 12.62$ eV and $V_3 = 1.50$ eV. Also five p electrons are transferred from the Cl and one s electron from the Na. We find a transfer energy of -108.44 and a repulsion of 48.10 eV for a cohesion of -60.33 compared with the experimental -16.91 eV, again an overestimate, but it is smaller than the cohesion for $MgSO_4$ which we calculated with the same bonding unit, and thus shows the appropriate experimental trend.

6A Dioxides

6A1 Sulphur Dioxide

We begin with sulphur dioxide, and shall return to carbon dioxide. We solve for the tight-binding energy of the SO_2 molecule without any approximations such as hybrids on the sulphur atom at the center, but bent as shown to the right in Fig. 6.3, and with an oxygen-sulphur distance of 1.43 Å. We again include only p states on the oxygen since the oxygen s state is so deep. We include the s state and the p states on the sulphur and we could include only the σ-oriented p state on the oxygen in the σ bonds, but it seems better to write the squared coupling between the sulphur pz state (see Fig. 6A.1) and the combination of oxygen p states as $2V_{pp\sigma}^2 \cos^2 \theta + 2V_{pp\pi}^2 \sin^2 \theta$ just as for the sulphate at the beginning of Appendix 5B. Then the squared coupling between the sulphur px state and the oxygen p states is $2V_{pp\sigma}^2 \sin^2 \theta + 2V_{pp\pi}^2 \cos^2 \theta$. Finally, as we saw in Table 3.1, our matrix elements from the free-electron fit needed to be scaled for molecules, 0.205 for O_2 and 0.346 for Cl_2, next to sulphur. We compromised and scaled here by 1/3. We can then, without further approximation, choose states even in reflection in the vertical axis, odd in that reflection, and π states perpendicular to the plane of the sketch in Fig. 6.A.1. The p states oriented along the y axis are simple; the odd combination of π states is not coupled to the sulphur states and has energy $\varepsilon_p(O)$. The even combination of oxygen states has coupling $V_2 = \sqrt{2}V_{pp\pi}$ to the sulphur π state and $V_3 = (\varepsilon_p(S) - \varepsilon_p(O))/2$ so the energies of the molecular orbitals are

$$\varepsilon_y = \frac{\varepsilon_p(S) + \varepsilon_p(O)}{2} \pm \sqrt{V_2^2 + V_3^2} \ . \tag{6A.1}$$

The σ states odd in a reflection through the vertical axis are also simple. Only the pz state on the sulphur enters, and it is coupled to the odd combination of the σ-oriented oxygen p states and to the odd combination of the π-oriented oxygen states. We again obtain Eq. (6A.1) with $V_2^2 = 2V_{pp\sigma}^2 \cos^2\theta + 2V_{pp\pi}^2 \sin^2\theta$ and with the same V_3 as for the π states. For the states even in reflection through the vertical axis the combination of oxygen

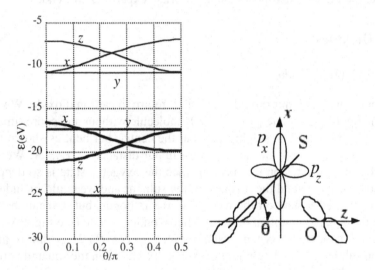

Fig. 6A.1. Energy levels for SO_2 as a function of the angle θ indicated in the sketch to the right.

$p\sigma$ states is coupled to the px sulphur orbital by $2V_{pp\sigma}{}^2\sin^2\theta + 2V_{pp\pi}{}^2\cos^2\theta$. We take it to be coupled to the sulphur s state by $\sqrt{2}V_{sp\sigma}$, though these three couplings would end up slightly different if we did the four-by-four problem with the oxygen $p\sigma$, $p\pi$, and sulphur pz and s. For the three-by-three problem, the corresponding three simultaneous equations lead to a three-by-three Hamiltonian matrix, with a secular equation,

$$(\varepsilon_s(S) - \varepsilon)(\varepsilon_p(S) - \varepsilon)(\varepsilon_p(O) - \varepsilon) - 2V_{sp\sigma}{}^2(\varepsilon_p(S) - \varepsilon) - 2V_{pp\sigma}{}^2\cos^2\theta(\varepsilon_s(S) - \varepsilon) = 0,$$

(6A.2)

which we can solve numerically. The plot of these levels as a function of θ/π is shown in Fig. 6A.1. Perpendicular orientation of the two bond axes occurs at $\theta = 0.25\pi$, in the middle and the values of larger θ have less meaning; at $\theta = 0.5\pi$ the two oxygen ions are at the same point. We have enough electrons to fill all but three levels (two electrons short of a full shell for each atom) and the occupied levels are shown heavy. We have not included nonorthogonality shifts, which change little with angle and the sum of the energies of occupied states is minimum at $\theta = \pi/4$, correctly predicting a bent molecule though with a bond angle just under 90° rather than the observed angle near 120°.

A very striking feature is the flatness of the lowest band, just below the sulphur s-state energy of −24.02 eV. The other states, and the total energy, would be little affected if we neglected its role, for example setting $V_{sp\sigma}$ equal to zero. Had we formed hybrids, the dangling hybrid would have had the sulphur s-state energy at $\theta = \pi/4$ and rise to the p state energy at $\theta = 0$ and $\pi/2$. It would have appeared to dominate the geometry, but metallization has eliminated almost all of its effect. For quantitative purposes hybrids are not a good idea here, as we suggested for diatomic molecules in Section 3.3. In fact a much better simplification is to use p states only, which again reduces the problem to quadratic equations.

If we set $V_{sp\sigma}$ equal to zero this low curve becomes completely flat and the middle curve based upon the p_x orbital drops down so that it equals the p_y curve at $\theta = 0$. The upper and middle x curves become mirror images of each other as do the upper and lower z curves. We proceed this way, and the problem becomes very simple. Neglecting $V_{sp\sigma}$ here makes almost no difference in the results, but we shall see that for the very similar carbon dioxide it would make a major difference.

With $V_{sp\sigma}$ taken as zero, the energy of the two σ-bond states at $\theta = \pi/4$, measured relative to the average of the two p-state energies at -14.19 eV, is given by $-\sqrt{(V_3^2 + V_{pp\sigma}^2)}$, with $V_3 = 2.59$ eV and $V_{pp\sigma} = 4.60$ eV with our 1/3 scaling. There is one π-bond state at $-\sqrt{(V_3^2 + 2V_{pp\pi}^2)}$ with the same V_3 and $V_{pp\pi} = -V_{pp\sigma}/3$. All of these are doubly occupied and if the molecules are taken apart one electron goes to each atom, with an average of the p-state energies, for each of these bonds. [There is also a nonbonding oxygen π state, with energy $\varepsilon_p(O)$ that would not change in the atomization.] The corresponding repulsions are added and, when minimized at the equilibrium spacing are equal to

$$\sum \frac{C}{d^4} = \frac{4\lambda V_{pp\sigma}^2}{\sqrt{\varepsilon_p(S)\varepsilon_p(O)}} + \frac{4\lambda V_{pp\pi}^2}{\sqrt{\varepsilon_p(S)\varepsilon_p(O)}} = \frac{2V_{pp\sigma}^2}{\sqrt{V_3^2 + V_{pp\sigma}^2}} + \frac{2V_{pp\pi}^2}{\sqrt{V_3^2 + 2V_{pp\pi}^2}}.$$

(6A.3)

We may add these up to obtain a total energy change of -18.43 eV, somewhat greater than the observed -11.13 eV. We may also solve Eq. (6A.3) for λ to obtain 1.46, comparable to other values we have obtained.

6A2 Carbon Dioxide

It is interesting to next consider the application to carbon dioxide, and can use the same sketch as to the right in Fig. 6A.1. The same orbitals enter and the most essential difference is having two less electrons. Recall that without the contribution of the central-atom s states in SO_2 the highest occupied x band had energy equal to that of the y band at $\theta = 0$ so that the y band, independent of angle, was the highest occupied band. Therefore, if the bands were the same for CO_2, and they will share such features, the only change, if $V_{sp\sigma}$ is set equal to zero, would be emptying this band independent of θ. We would find the minimum energy with perpendicular bonds, though CO_2 in fact is straight, $\theta = 0$. However, if we include the effect of the central-atom s states we see in Fig. 6A.1 that at very small θ the x band rises above and becomes the empty band, as redrawn in Fig. 6A.2, where the total energy from that plot is also shown as a light line. Then emptying the highest band modifies the shape of the total energy as shown by the heavy line. Indeed $\theta = 0$ structure then has the lowest energy as in experiment.

Thus the difference in geometry between bent SiO_2 and the straight CO_2 arises directly from the central atom s state and whether this highest state is occupied. In that sense the argument based upon hybrids, with two forming bonds and the dangling hybrid above them, was qualitatively correct, and we shall explore that further in Appendix 6B. If the dangling hybrid is occupied, the molecule is bent; if empty, it is straight. On the other hand, the energy difference between the two states is on the order of two volts, rather than the $\varepsilon_p - \varepsilon_s$ of the central atom, 8 eV for carbon and 12 eV for sulphur. Further, which minimum in the total energy in Fig. 6A.2 is lower is a quantitative question depending upon the parameters of the system. Hybrids can thus be useful for qualitative arguments for these molecules, but quantitatively they are very questionable, as will become particularly clear in Appendix 6B.

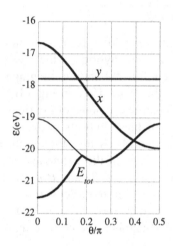

Fig. 6A.2. The crossing levels, x, and y, from Fig. 6A.1, and the total energy as a light line for SO_2, for which both bands are doubly occupied. CO_2 has one band empty, and if it were the y band, the total energy would be the same, but shifted by a constant. Correctly leaving the highest band empty changes the shape of the total-energy curve as shown by the heavy line.

6B Dioxides with Hybrids

In order to see more clearly the quantitative difficulties with the use of hybrid states for these molecules we make application to the SO_2 treated in Appendix 6A. We use the same set of orbitals and couplings which we used in solving the full tight-binding problem there, and again reduce our coupling by a factor 1/3, but now use hybrid states.

The formation of sulphur hybrids is illustrated in Fig. 6B.1. In constructing sp hybrids, we should keep them orthogonal to each other, requiring

$$|h_{\pm}> = \frac{\sqrt{\cos^2\theta - \sin^2\theta}\,|s> + |p\sigma>}{\sqrt{2}\cos\theta} \tag{6B.1}$$

with the state $|p\sigma>$ oriented along each of the internuclear distances. Then there is a third hybrid, a dangling hybrid, orthogonal to both of these with the px state vertical, given by

$$|h_d> = \frac{\sin\theta\,|s> + \sqrt{\cos^2\theta - \sin^2\theta}\,|px>}{\cos\theta}. \tag{6B.2}$$

A special case is with $\theta = 30°$. Then all three are of the form $(|s> + \sqrt{2}|p>)/\sqrt{3}$. These are sp^2 hybrids, which we postulated for benzene and which could be used for σ-bonding in graphite, leaving a π state oriented perpendicular to the plane. Here in Fig. 6B.1 there is also a sulphur p state oriented perpendicular to the plane of the figure.

An important feature of these hybrids is that if the angles change, the energies of the individual hybrids change, though the average hybrid energy remains the same as the average of the s state and the two in-plane p states. Those in Eq. (6B.1) are $\varepsilon_h = (\cos^2\theta - \sin^2\theta)\varepsilon_s + \varepsilon_p)/(2\cos^2\theta)$ and in Eq. (6B.2) $\varepsilon_d = (\sin^2\theta\varepsilon_s + (\cos^2\theta - \sin^2\theta)\varepsilon_p)/\cos^2\theta$.

For any value of θ we may form these hybrids, construct bonding and antibonding states for each neighbor, including only the coupling

$$V_2 = \frac{\sqrt{\cos^2\theta - \sin^2\theta}\,V_{sp\sigma} + V_{pp\sigma}}{\sqrt{2}\cos\theta} \tag{6B.3}$$

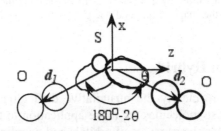

Fig. 6B.1. An SO_2 molecule, with hybrids formed on the central sulphur, coupled to the σ-oriented p states on the oxygen atoms. $d_1 = d_2 = 1.43$ Å. $\theta \approx 30°$.

between the hybrid and the p state toward which it is oriented, with a $V_3 = (\varepsilon_h - \varepsilon_p(O))/2$, just as we did for the tetrahedral semiconductors. These lead to bonding and antibonding energies

$$\varepsilon = (\varepsilon_h + \varepsilon_p(O))/2 \pm \sqrt{(V_2^2 + V_3)} \qquad (6B.4)$$

and π-bonding energies of the same form as in Eq. (6A.1).

$$\varepsilon = (\varepsilon_p(S) + \varepsilon_p(O))/2 \pm \sqrt{((\varepsilon_p(S) - \varepsilon_p(O)))^2/4 + 2V_{pp\pi}^2}. \qquad (6B.5)$$

Independent bonds have been formed with each of the oxygen atoms and the corresponding bond and antibond levels, both doubly degenerate, are shown in Part (b) of Fig. 6B.2, along with the energy of the dangling hybrid. In Part (a) are shown the corresponding levels from the full calculation, the x and z levels of Fig. (6A.1). The figures stop at $\theta = \pi/4$ since orthogonal hybrids cannot be constructed for angles between the bonds of less than $\pi - 2\theta = \pi/2$ between them.

The two sets of curves are so different the it is not easy to compare them. We could take an even combination of the two bond states (or antibond states) and they could correspond to the x levels and an odd combination would correspond to the z levels, without changing the figure.

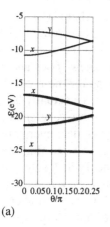

(a)

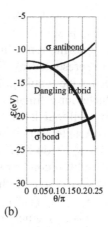

(b)

Fig. 6B.2. In Part (a) are shown the x and z energy levels from Fig. 6A.1, the full calculation. In Part (b) are the corresponding levels based upon hybrids, without any metallization included. The bonds and antibonds are doubly degenerate.

Then adding the many couplings we have neglected in the hybrid representation would make Part (b) become the same as Part (a). The largest effect is the coupling of the dangling hybrid to the even combination of bond and antibond states. These deform the dangling hybrid into the central x state in Part (a) shifting the even combinations of bonds and antibonds into the x levels in Part (a). There are also couplings between the even combinations of bonds and between the odd combinations of bonds, and similar couplings for the odd combinations. It hardly seems appropriate to try to make these corrections, rather than simply doing the full tight-binding calculation. In spite of these difficulties, we found it useful in the main text to see how hybrid states give qualitatively correct geometries in a wide range of molecules.

8A The Special-Points Method

The simplest case to treat is energy bands in rock salt, with one sodium s state coupled to the three p states on its six chlorine neighbors. The energy bands will be given by

$$\varepsilon_k = \frac{\varepsilon_s + \varepsilon_p}{2} \pm \sqrt{\left(\frac{\varepsilon_s - \varepsilon_p}{2}\right)^2 + V_{sp\sigma}^2 f(\mathbf{k})}, \tag{8A.1}$$

similar to the pd bands along a [100] direction, where f($\mathbf{k}$) was $4\sin^2(kd)$. For these sp bands at every wavenumber there are two combinations of the p states on the chlorine ions which are uncoupled to the s state and the other combination represented by Eq. (8A.1). If we knew $f(\mathbf{k})$ and the special wavenumber $\mathbf{k}^*$ we would know what $f(\mathbf{k}^*)$ was for the rock-salt structure and would have our estimate of the band energy. We actually have a way of getting $f(\mathbf{k}^*)$ without knowing either. If $V_{sp\sigma}$ were very small, we could expand Eq. (8A.1) to second order, giving an upper-state energy of $\varepsilon_{\mathbf{k}^*} = \varepsilon_s + V_{sp\sigma}^2 f(\mathbf{k}^*)/(\varepsilon_s - \varepsilon_p)$ However, for very small $V_{sp\sigma}$ we can find the energy of that upper state by second-order perturbation theory as a sum of shifts from the coupling with each of the $X = 6$ neighbors as $= \varepsilon_s + X V_{sp\sigma}^2/(\varepsilon_s - \varepsilon_p)$. Thus we know that $f(\mathbf{k}^*)$ should be equal to the number of nearest neighbors, X. It is convenient to write the square root in Eq. (8A.1) as

$$\varepsilon_{k^*} = \pm\sqrt{V_3^2 + V_2^2} \tag{8A.2}$$

measured from the average atomic-level energy, $(\varepsilon_s(\text{Na}) + \varepsilon_p(\text{Cl}))/2$, with $V_3 = (\varepsilon_s(\text{Na}) - \varepsilon_p(\text{Cl}))/2$, and V_2^2 in this case $6V_{sp\sigma}^2$.

Note that we needed to base this on the shift of the s state since we did not know what combination of p states to take to do the perturbation theory. This uncertainty comes up again when we seek to generalize this to pd bands in the rock-salt structure. We should focus on the three p states, not the five d states, and in perturbation theory each of the three p states can be seen to be shifted by $(2V_{pd\sigma}^2 + 4V_{pd\pi}^2)/(\varepsilon_p - \varepsilon_d)$. There would not be any $\mathbf{k}$ at which all three bands had this same energy, but we can imagine a $\mathbf{k}^*$ at which one band was shifted by $6V_{pd\sigma}^2/(\varepsilon_p - \varepsilon_d)$ and two bands shifted by $6V_{pd\pi}^2/(\varepsilon_p - \varepsilon_d)$. This would lead again to Eq. (8A.1) with $f(\mathbf{k}^*) = X$, the number of nearest neighbors, with one band with $V_2^2 = 6V_{pd\sigma}^2$ and two bands with $V_2^2 = 6V_{pd\pi}^2$ for Eq. (8A.2) This is not really a derivation, unless in a Heaviside-algebra sort of way, but seems a plausible way to proceed.

A similar problem arises in the perovskite structure of Fig. 8.3. There are then five transition-metal d states and nine oxygen p states per formula unit (and we leave out the A-site states for this analysis). There would be five upper and five lower bands and four oxygen nonbonding p bands. In this case the t_g states, xy, yz, and zx are coupled only to π-oriented p states and can be treated separately as three bands with $V_2^2 = 4V_{pd\pi}^2$. Then the e_g states lead to two bands with $V_2^2 = 3V_{pd\sigma}^2$.

A remarkable feature of this result for the perovskite structure is that these correspond to the cluster orbitals used in Harrison (2008, 2009, and 2010) and here in Section 8.5. Thus we were using the same formulae whether we were thinking of energy bands or thinking of states localized to the individual manganese ions, coupled to the states of their oxygen neighbors. This was not true for MgO, in the rock-salt structure, where the manganese cluster orbitals would be the same as for the perovskites, but the bands would need to be based on the oxygen p states, as described above.

References

Andersen, O. K., 1973, Solid State Commun. **13**, 133. *108*

Andersen, O. K., and O. Jepsen, 1977, Physica (Utrecht) **91B**, 317. *108*

Andersen, O. K., W. Klose, and H. Nohl, 1978, Phys. Rev. B**17**, 1209. *107*

Anderson, Alfred B., and Roald Hoffmann, 1974, J. Chem. Phys. **60**, 4271. *156*

Anderson, P. W., 1950, Phys. Rev. **79**, 350. *129*

Ashccroft, N. W, 1966, Phys. Lett. **43**, 48. *8*

Bader, Richard. F. W., 1970, *Atoms and Molecules*, www.chemistry.mcmaster.ca/esam/Chapter_7/section_2.html.

Baker, A. D. and M. D. Baker, 2009, *Madelung constants of nanoparticles and nanosurfaces*, J. Phys. Chem. C, **113**, 14793 and 2010, *Rapid calculation of individual ion Madelung constants and their convergence to bulk values*, American J. Phys. **78**, 102. *83, 90*

Born, M., 1931, Ergebn. d. exakt. Naturw. **10**, 387. *88*

Castner, T. G., and W. Känzig, 1957, J. Phys. Chem. Solids **3** 178. *94*

Chelikowsky, James R., and Marvin L. Cohen, 1976, *Nonlocal Pseudopotential Calculations for the Electronic Structure of Eleven Diamond and Zinc-Blende Semiconductors*, Phys. Rev. B **14**, 556. *75*

Emin, D., and T. Holstein, 1969, Ann. Phys. (NY) **53**, 439. See also David Emin, 1998, *Polaron formation and motion in magnetic*, in Science and Technology of Magnetic Oxides, M. F. Hundley, J. H. Nickel, and R. Ramesh and Y. Tokura, eds. Materials Research Society Symposium Proceedings Vol. 494 (Materials Research Society, Pittsburgh,), pp. 163-174. *134*

Feibelman, Peter J., 2010, *The first wetting layer on a solid*, Physics Today **63**, 34. *30ff, 67, 152,153, 173*

Friedel, J., 1969, in *The Physics of Metals*, ed. by J. M. Ziman, Cambridge University Press, New York. *81, 110ff, 112*

Froyen, Sverre, and W. A. Harrison, 1979, *Elementary Prediction of LCAO Matrix Elements*, Phys. Rev. B **20**, 2420. *108*

Gillespie, R. J., 1972, *Molecular Geometry*, Van Nostrand (London). *102, 103, 175*

Harrison, W. A., 1979, *Theory of Polar Semiconductor Surfaces*, J. Vac. Sci. Tech. **16**, 1492.

Harrison, W. A., 1981, *New tight-binding parameters for covalent solids obtained using Louie peripheral states*, Phys. Rev. B **24**, 5835 . *75, 136*

Harrison, W. A. , 1989, *Electronic Structure and the Properties of Solids*, W. H. Freeman, (San Francisco, 1980); reprinted by Dover (New York, 1989).

Harrison, W. A. 1999, *Elementary Electronic Structure*, World Scientific, Singapore, revised edition (2004).

Harrison, W. A., 2000, *Applied Quantum Mechanics*, World Scientific, Singapore.

Harrison, W. A., 2002, *p-, d-, and f-Bonds in Elemental Solids*, Philosophical Magazine B **82**, 1755. *77ff*

Harrison, W. A., 2003, *Tight-binding theory of surface states in metals*, Physica Scripta **67**, 253. *65*

Harrison, W. A., 2004a, *Understanding phonons in plutonium*, Phys. Rev. B **69**, 113106. *121*

Harrison, W. A., 2004b, *Thermal properties of plutonium*, Phys. Rev. B **69**, 224109. *121*

Harrison, W. A., 2006a, *Simple calculation of Madelung constants*, Phys. Rev. B **73**, 212103. *88*

Harrison, W. A., 2006b, *Tight-binding theory of the dielectric susceptibility and transverse charge of insulators*, Phys. Rev. B **74**, 205101. *11, 93*

Harrison, W. A., 2006c, *Valence-skipping compounds as positive-U electronic systems*, Phys. Rev. B **74**, 245128. *117*

Harrison, W. A., 2007, *Heisenberg exchange in the magnetic monoxides*. Phys. Rev. B **76**, 054417. *108, 129*

Harrison, W. A., 2008, *Tight-binding theory of the oxides of manganese and iron*, Phys. Rev. B77, 245103 (arXiv:0803.0994). *127, 185*

Harrison, W. A., 2009, *Tight-binding theory of lanthanum strontium manganate*, arXiv:0807.2248, submitted to Phys. Rev., but not accepted. *132ff, 185*

Harrison, W. A., 2010, *Oxygen atoms and molecules at $La_{1-x}Sr_xMnO_3$ surfaces*, Phys. Rev. B **81**, 045433 (arXiv:0911.2268). *137ff, 172, 185*

Harrison, W. A. , and Sverre Froyen, 1980 , *Universal LCAO Parameters for d-State Solids*, Phys. Rev. B **21**, 3214. *109*

Harrison, W. A., and Galen K. Straub, 1987, *Electronic structure and bonding in d- and f-metal AB compounds*, Phys. Rev. B **35**, 2695. *87*

Hartree, D. R., 1928, *Proc. Cambridge Phil. Soc.* **24**, 89. *1, 34*

Hoffmann, Roald, 1963, *An Extended Hückel Theory. I. Hydrocarbons*, J. Chem. Phys. **39**, 1397. *33, 35, 156*

Holm, U., and K. Kerle, 1990, Mol. Phys. **69**, 803. *12*

Irvine, J. M., 1972, *Nuclear Structure Theory*, Pergamom Press, Oxford. *12*

Kittel, C., 1976, *Introduction to Solid State Physics*, 5th ed., Wiley, New York, (or 3rd ed., 1966).

Kotomin, E. A., Y. A. Mastrikov, E. Heifets, and J. Maier, 2008, *Adsorption of atomic and molecular oxygen on the LaMnO$_3$(001) surface:* ab initio *supercell calculations and thermodynamics*, Phys. Chem. Chem. Phys. **10**, 4644. *140*

Lee, Yueh-Lin, Jesper Kleis, Jan Rossmeisl, and Dane Morgan, 2009, *ab initio energetics of LaBO$_3$(001) (B=Mn, Fe, Co and Ni) for solid oxide fuel cell cathodes* Phys. Rev. B **80**, 224101. *137, 140*

Mannhart, J., D. H. A. Blank, H. Y. Hwang, A. J. Millis, and J. M. Triscone, 2008, MRS Bull. **33**, 1027. *136*

Mann, J. B., 1967, *Atomic Structure Calculations, 1: Hartree-Fock Energy Results for Elements Hydrogen to Lawrencium..* Distributed by Clearinghouse for Technical Information, Springfield, Virginia 22151. *2, 3, 7, 104*

Mattheiss, L. F., 1972, Phys. Rev. **B6**, 4718. *131*

Pantelides, S. T., and W. A. Harrison, 1975, *The Structure of the Valence Bands of the Zincblende-Type Semiconductors*, Phys. Rev. B **11**, 3006. *4*

Pantelides S. T., and W. A. Harrison, 1976 *Electronic Structure, Spectra and Properties of Four-Two Coordinated Materials I Silicon Dioxide and Germanium Dioxide*, Phys. Rev. **B** 13, 2667. *82ff, 173ff*

Pauling, Linus, 1960, *The Nature of the Chemical Bond*, Cornell University Press, Ithaca, New York. *45, 79*

Perdew, J. P., and S. H. Vosko, 1974, J. Phys. F, **4**, 380. *167, 169*

Rømark, L., S. Stølen, K. Wiik, and T. Grande, 2002, J. Solid State Chem. **163**, 186. *133*

Rosenfeld, M, M Ziegler, and W. Kanzig, 1978, *Xray Study of the Low Temperature Phases of Alkali Hyperoxides*, Helvetia Physica Acta **51**, 299. *95*

Ruderman, M. A., and C. Kittel, 1954, Phys. Rev. **96**, 99. *116*

Shockley, W., 1939, Phys. Rev. **56**, 317. *65*

Slater, J. C., and G. F. Koster, 1954, Phys. Rev. **94**, 1498. *106, 127*

Straub, G. K., and W. A. Harrison, 1985, *Analytic methods for calculation of electronic structure of solids*, Phys. Rev. B **31**, 7668, described also in Harrison (1999), 556ff. *108*

Tai, L.-W., M. M. Nasrallah, H. U. Anderson, D. M Sparlin, and S. R. Schlin, 1995, Solid State Ionics, **76**, 273. *134*

Tamm, I., 1932, Z. Phys. **76**, 849.

Weast, R. C., 1975: ed., *Handbook of Chemistry and Physics*, 56th Edition, The Chemical Rubber Company, Cleveland.

Wills, J. M., and W. A. Harrison, 1983, *Interionic Interactions in Transition Metals*, Phys. Rev. B **28**, 4363. *113*

Zhou, J. S., and J. B. Goodenough, 2002, Phys. Rev. Lett. **89**, 87201. *135*

Subject Index